T0240718

# SpringerBriefs in Mathematics

SpringerBriefs present concise summaries of cutting-edge research and practical applications across a wide spectrum of fields. Featuring compact volumes of 50 to 125 pages, the series covers a range of content from professional to academic. Briefs are characterized by fast, global electronic dissemination, standard publishing contracts, standardized manuscript preparation and formatting guidelines, and expedited production schedules.

**Typical topics might include:**

- A timely report of state-of-the art techniques
- A bridge between new research results, as published in journal articles, and a contextual literature review
- A snapshot of a hot or emerging topic
- An in-depth case study
- A presentation of core concepts that students must understand in order to make independent contributions

**SpringerBriefs in Mathematics** showcases expositions in all areas of mathematics and applied mathematics. Manuscripts presenting new results or a single new result in a classical field, new field, or an emerging topic, applications, or bridges between new results and already published works, are encouraged. The series is intended for mathematicians and applied mathematicians. All works are peer-reviewed to meet the highest standards of scientific literature.

*Titles from this series are indexed by Scopus, Web of Science, Mathematical Reviews, and zbMATH.*

Stasys Jukna

# Tropical Circuit Complexity

## Limits of Pure Dynamic Programming

 Springer

Stasys Jukna (iD)
Faculty of Mathematics and Informatics
Vilnius University
Vilnius, Lithuania

ISSN 2191-8198          ISSN 2191-8201    (electronic)
SpringerBriefs in Mathematics
ISBN 978-3-031-42353-6          ISBN 978-3-031-42354-3    (eBook)
https://doi.org/10.1007/978-3-031-42354-3

Mathematics Subject Classification: 49L20; 68Q06; 68Q17; 90C39

This Springer imprint is published by the registered company Springer Nature Switzerland AG
The registered company address is: Gewerbestrasse 11, 6330 Cham, Switzerland

Paper in this product is recyclable.

# Preface

> *Go to the roots of calculations! Group the operations. Classify them according to their complexities rather than their appearances! This, I believe, is the mission of future mathematicians.*
>
> *–Evariste Galois*

Understanding the power and weakness of algorithmic paradigms for solving decision or optimization problems in rigorous mathematical terms is an important long-term goal. Along with greedy and linear programming, dynamic programming (DP) is one of THE algorithmic paradigms for solving combinatorial optimization problems. Dynamic programming algorithms turned out to be quite powerful in many practical applications, so that we know what these algorithms *can* do. But what can DP algorithms *not* do (efficiently)? Answering this question is the subject of this book.

Roughly speaking, the idea of DP is to break up a given optimization problem into smaller subproblems in a divide-and-conquer manner and solve these subproblems recursively. Optimal solutions of smaller instances are found and retained for use in solving larger instances (smaller instances are never solved again). Many classical DP algorithms are *pure* in that they only apply the basic operations $(\min, +)$ or $(\max, +)$ in their recursion equations.

A rigorous mathematical model for pure DP algorithms is that of tropical circuits. These are conventional combinational circuits using $(\min, +)$ or $(\max, +)$ operations as gates. Pure DP algorithms are special (recursively constructed) tropical circuits. So, if one can prove that *any* tropical circuit solving a given optimization problem must use at least $t$ gates, then we know that *no* pure DP algorithm can solve this problem by performing fewer than $t$ $(\min, +)$ or $(\max, +)$ operations, be the designer of an algorithm even omnipotent. Thanks to the rigorous combinatorial nature of tropical circuits, ideas and arguments from the Boolean and arithmetic circuit complexity can be exploited to obtain lower bounds for topical circuits and, hence, also for pure DP algorithms.

For example, the classical Bellman–Held–Karp DP algorithm gives a tropical $(\min, +)$ circuit with about $n^2 2^n$ gates solving the travelling salesman problem on $n$-vertex graphs, while a trivial brute force algorithm results in about $n! \approx (n/e)^n$

gates. On the other hand, Jerrum and Snir in 1982 have shown that at least about $n^2 2^n$ gates are also necessary in any (min, +) circuit solving this problem. This shows that the Bellman–Held–Karp DP algorithm is *optimal* among all pure DP algorithms for this problem. The tropical (min, +) circuit corresponding to the (also classical) Floyd–Warshall–Roy pure DP algorithm for the all-pairs shortest paths problem on $n$-vertex graphs uses about $n^3$ gates. On the other hand, already in 1970, Kerr has shown that at least about $n^3$ gates are also necessary for this problem. So, the Floyd–Warshall–Roy pure DP algorithm is also optimal in the class of all pure DP algorithms.

After these and several other impressing lower bounds where obtained, a long break followed. Only in recent years, and mainly due to recognized connection with dynamic programming, tropical circuits have attracted growing attention again. The goal of this book is to survey the lower-bound ideas and methods that emerged during these last years.

We focus on presenting the lower-bound arguments themselves, rather than on quantitative bounds achieved using them. That is, the focus is on the proof arguments, on the ideas behind them. Because of a very pragmatic motivation of tropical circuits—their intimate relation to dynamic programming—the primary goal is to create as large as possible "toolbox" for proving lower bounds on the size of tropical circuits, not relying on unproven complexity assumptions like $\mathbf{P} \neq \mathbf{NP}$.

The difficulty in proving that a given optimization problem requires large tropical circuits lies in the nature of our adversary: the circuit. Small circuits may work in a counterintuitive fashion, using deep, devious, and fiendishly clever ideas. How can one prove that there is no clever way to quickly solve the problem? In this book, we will learn some tools to defeat this adversary.

Tropical algebra and geometry—where "adding" numbers means to take their minimum or maximum, and "multiplying" them means to add them—are now actively studied topics in mathematics. Tropical circuit complexity adds a computational complexity aspect to this topic.

The book is self-contained and is meant to be approachable already by graduate students in mathematics and computer science. The text assumes certain mathematical maturity (minor knowledge of basic concepts in graph theory, discrete probability, and linear algebra) but *no* special knowledge in the theory of computing or dynamic programming.

Supplementary material to the book can be found on my home page.

Vilnius, Lithuania/Frankfurt, Germany                                    Stasys Jukna
June 2023

# Contents

# Notation

We will use more or less standard concepts and notation. For ease of reference, let us collect some of most often used ones right now:

| | |
|---|---|
| Nonnegative real numbers | $\mathbb{R}_+ = \{x \in \mathbb{R} : x \geqslant 0\}$ |
| Nonnegative integers | $\mathbb{N} = \{0, 1, 2, \ldots\}$ and $[n] = \{1, \ldots, n\}$ |
| $K_n$ | The complete graph on $[n]$ |
| $K_{n,n}$ | A complete bipartite $n \times n$ graph |
| $2^X$ for a set $X$ | Family of all subsets of $X$ |
| $|X|$ for a finite set $X$ | Number of elements in $X$ |
| Family $\mathcal{F} \subseteq 2^X$ is uniform | All sets in $\mathcal{F}$ have the same cardinality |
| Characteristic vector of $S \subseteq [n]$ | Vector $a \in \{0, 1\}^n$ with $a_i = 1$ iff $i \in S$ |
| Unit vector $\vec{e}_i$ | $\vec{e}_i = (0, \ldots, 0, 1, 0, \ldots, 0)$ with 1 in the $i$th position |
| $a \leqslant b$ for $a, b \in \mathbb{R}^n$ | $a_i \leqslant b_i$ for all $i = 1, \ldots, n$ |
| $A \subseteq \mathbb{R}^n$ is an antichain | $a \not\leqslant b$ for all $a \neq b \in A$ |
| Upward closure $A^\uparrow$ of $A \subseteq \mathbb{N}^n$ | $A^\uparrow = \{b \in \mathbb{N}^n : b \geqslant a$ for some $a \in A\}$ |
| Downward closure $A^\downarrow$ of $A \subseteq \mathbb{N}^n$ | $A^\downarrow = \{b \in \mathbb{N}^n : b \leqslant a$ for some $a \in A\}$ |
| $B \subseteq \mathbb{R}^n$ lies above $A \subseteq \mathbb{R}^n$ | $B \subseteq A^\uparrow$, i.e., $\forall b \in B\ \exists a \in A : b \geqslant a$ |
| $B \subseteq \mathbb{R}^n$ lies below $A \subseteq \mathbb{R}^n$ | $B \subseteq A^\downarrow$, i.e., $\forall b \in B\ \exists a \in A : b \leqslant a$ |
| Support of $a \in \mathbb{R}^n$ | $\sup(a) = \{i : a_i \neq 0\}$ |
| Degree of $a \in \mathbb{N}^n$ | $|a| = a_1 + \cdots + a_n$ |
| Lower envelope of $A \subseteq \mathbb{N}^n$ | $\lfloor A \rfloor = \{a \in A : |a|$ is minimal$\}$ |
| Higher envelope of $A \subseteq \mathbb{N}^n$ | $\lceil A \rceil = \{a \in A : |a|$ is maximal$\}$ |
| $A \subseteq \mathbb{N}^n$ is homogeneous | $\lceil A \rceil = \lfloor A \rfloor$ |
| Sum of $a, b \in \mathbb{R}^n$ | $a + b = (a_1 + b_1, \ldots, a_n + b_n)$ |
| Minkowski sum of $A, B \subseteq \mathbb{R}^n$ | $A + B = \{a + b : a \in A, b \in B\}$ |
| Scalar product of $a, b \in \mathbb{R}^n$ | $\langle a, b \rangle = a_1 b_1 + \cdots + a_n b_n$ |
| Tropical (min, +) polynomial | $f(x) = \min_{a \in A}\{\langle a, x \rangle + c_a\}; A \subseteq \mathbb{N}^n, c_a \in \mathbb{R}_+$ |

# Chapter 1
# Basics

**Abstract** In this chapter, we recall the models of arithmetic $(+, \times)$, Boolean $(\vee, \wedge)$, and tropical $(\min, +)$ and $(\max, +)$ circuits, introduce Minkowski $(\cup, +)$ circuits as a model taking all of them "under one hat," establish the basic structural properties of tropical polynomials, and relate the corresponding circuit complexity measures. The main message of this chapter is that: lower bounds on the tropical circuit complexity of optimization problems can be obtained by proving lower bounds on the monotone arithmetic circuit complexity of particular polynomials.

## 1.1 What Is This Book About?

We are interested in solving discrete optimization problems[1] $f : \mathbb{R}_+^n \to \mathbb{R}_+$ on given sets $A \subseteq \mathbb{N}^n$ of *feasible solutions*:

$$f(x) = \min_{a \in A} \sum_{i=1}^{n} a_i x_i \quad \text{or} \quad f(x) = \max_{a \in A} \sum_{i=1}^{n} a_i x_i . \tag{1.1}$$

If $A \subseteq \{0, 1\}^n$, then $f$ is usually called a *combinatorial optimization* or $0/1$ *optimization* problem. The set $A \subseteq \mathbb{N}^n$ of feasible solutions can be described either explicitly (as a particular set of vectors), or as the set $A = \{a \in \mathbb{N}^n : Ma \leqslant b\}$ of nonnegative integer or $0/1$ solutions of a given system of linear inequalities (as in linear programming), or by other means. It is only important that the set $A$ does not depend on the input weightings $x \in \mathbb{R}_+^n$.

For example, in the $0/1$ optimization problem, known as the shortest $s$-$t$ path problem on a given graph $G$, the set $A$ of feasible solutions consists of characteristic $0$-$1$ vectors[2] of all paths in $G$ between two vertices $s$ and $t$, the paths being viewed as sets of their edges. In the minimum weight spanning tree problem, feasible solutions

---

[1] In what follows, $\mathbb{N} = \{0, 1, 2, \ldots\}$ stands for the set of all nonnegative integers, $[n] = \{1, \ldots, n\}$ for the set of the first $n$ positive integers, and $\mathbb{R}_+$ for the set of all nonnegative real numbers.

[2] The *characteristic* $0$-$1$ *vector* of a set $S \subseteq [n]$ is the vector $a \in \{0, 1\}^n$ such that $a_i = 1$ iff $i \in S$.

© The Author(s), under exclusive license to Springer Nature Switzerland AG 2023
S. Jukna, *Tropical Circuit Complexity*, SpringerBriefs in Mathematics,
https://doi.org/10.1007/978-3-031-42354-3_1

are (characteristic 0-1 vectors of) spanning trees of a given graph, and the problem is to compute the minimum weight of such a tree. In the assignment problem, we deal with perfect matchings in a complete bipartite graph, etc.

Note that min, max, and $+$ are the only operations used to formulate the problems (1.1) themselves. It is easy to see that the optimization problem on *every* set $A \subseteq \{0, 1\}^n$ of feasible solutions *can* be solved using at most $n|A|$ (min, $+$) or (max, $+$) operations: compute all $|A|$ sums $\sum_{i=1}^{n} a_i x_i$ for $a \in A$, and take their minimum or maximum using additional $|A| - 1$ min or max operations. But, as Examples 1.6 to 1.8 in Sect. 1.4 show, this trivial *upper* bound can be very far from the truth. For example, if $A \subseteq \{0, 1\}^n$ consists of all $|A| = \binom{n}{n/2}$ vectors with exactly $n/2$ ones, then the minimization problem on $A$ is, given an input weighting $x \in \mathbb{R}^n_+$, to compute the sum of lightest $n/2$ weights. Although there are $|A| \geqslant 2^{n/2}$ feasible solutions, the problem can be solved by using at most $\mathcal{O}(n^2)$ (min, $+$) operations (Example 1.6). The main goal of this book is to learn how to prove *lower* bounds:

> (∗)    *At least* how many (min, $+$) or (max, $+$) operations do we need to solve or to approximate a given discrete optimization problem?

For example, in the *lightest triangle problem*, inputs are assignments of nonnegative real weights to the edges of the complete graph $K_n$ on $\{1, \ldots, n\}$ and the goal is to compute the minimum weight of a triangle. How many (min, $+$) operations do we need to solve this problem (for all possible input weightings)? Since there are only $\binom{n}{3}$ triangles, the problem can be trivially solved using a cubic number $\mathcal{O}(n^3)$ of (min, $+$) operations. More interesting, however, is the question: does a cubic number $\Omega(n^3)$ of operations is also *necessary* to solve this problem? (Yes, Corollary 2.4.)

With a wish to make the question (∗) mathematically precise, we arrive to the classical model of *circuits* (also called *combinational circuits*). We will mainly be interested in tropical circuits, that is, in circuits over tropical[3] semirings $(\mathbb{R}_+, \min, +)$ and $(\mathbb{R}_+, \max, +)$. But, as we will see, the power of tropical circuits is related to that of circuits over the Boolean semiring $(\{0, 1\}, \vee, \wedge)$ as well as over the arithmetic semiring $(\mathbb{R}_+, +, \times)$. So, let us first recall what "circuits" over a semiring actually are.

---

[3] The adjective "tropical" is not to contrast with "polar geometry." It was coined by French mathematicians in honor of Imre Simon who lived in Sao Paulo (south tropic). Tropical algebra and tropical geometry are now intensively studied topics in mathematics.

## 1.2 Circuits

A (commutative) *semiring* $(R, \oplus, \odot)$ consists of a set $R$ closed under two associative and commutative binary operations "addition" $x \oplus y$ and "multiplication" $x \odot y$, where multiplication distributes over addition: $x \odot (y \oplus z) = (x \odot y) \oplus (x \odot z)$. That is, in a semiring, we can "add" and "multiply" elements, but neither "subtraction" nor "division" is necessarily possible. A semiring is *additively idempotent* if $x \oplus x = x$ holds and is *multiplicatively idempotent* if $x \odot x = x$ holds for all elements $x \in R$. A semiring may (or may not) contain an additive neutral element $0 \in R$ satisfying $0 \oplus x = x \oplus 0 = x$. We will only assume that the semiring contains a multiplicative neutral element $1 \in R$ such that $1 \odot x = x \odot 1 = x$.

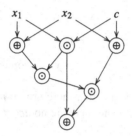

A *circuit* $\Phi$ (also known as a *combinational circuit*) over a semiring $(R, \oplus, \odot)$ is a directed acyclic graph; parallel edges joining the same pair of nodes are allowed. Each indegree-zero node (a *source* node) holds either one of the variables $x_1, \ldots, x_n$ or a semiring element $c \in R$; if there are no semiring elements $c \in R$ other than $c = 1$ as inputs, then the circuit is called *constant-free*. Every other node, a *gate*, has indegree two and performs one of the semiring operations $\oplus$ or $\odot$ on the values computed at the two gates entering this gate. Usually (but not always), one of the gates is declared as the output gate. The *size* of a circuit $\Phi$, denoted size($\Phi$), is the total number of gates in it. A circuit $\Phi$ *computes* a function $f : R^n \to R$ if $\Phi(x) = f(x)$ holds for all $x \in R^n$.

**Proposition 1.1** *Over any semiring, there are at most $2^s(2s + n + 1)^{2s}$ distinct constant-free circuits $\Phi(x_1, \ldots, x_n)$ with at most $s$ gates.*

**Proof** Each gate in such a circuit is assigned a semiring operation (two choices) and acts on some two previous nodes. Each previous node can be either a previous gate (at most $s$ choices) or an input variable ($n$ choices) or the "constant" $1$. Thus, each single gate has at most $N = 2(2s + n + 1)^2$ choices, and the number of choices for a circuit is at most $N^s$. □

In this book, we will consider circuits over the following four semirings $(R, \oplus, \odot)$: the arithmetic semiring $(\mathbb{R}_+, +, \times)$ with usual (arithmetic) addition and multiplication, the tropical (min, +) semiring $(\mathbb{R}_+, \min, +)$, the tropical (max, +) semiring $(\mathbb{R}_+, \max, +)$, and the Boolean $(\vee, \wedge)$ semiring $(\{0, 1\}, \vee, \wedge)$. That is, we will consider the following types of circuits:

**Table 1.1** Properties making tropical circuits an *intermediate* model between monotone Boolean and monotone arithmetic circuits. We will see (as summarized in Sect. 1.13) that the power of tropical circuits lies *properly* between that of monotone Boolean and monotone arithmetic circuits

| Semiring $(\oplus, \odot)$ | $\oplus$-Idempotence $x \oplus x = x$ | $\odot$-Idempotence $x \odot x = x$ | Absorption $x \oplus (x \odot y) = x$ |
|---|---|---|---|
| Boolean $(\vee, \wedge)$ | + | + | + |
| Tropical $(\min, +)$ | + | − | + |
| Arithmetic $(+, \times)$ | − | − | − |

$x \oplus y := x + y$ and $x \odot y := xy$        monotone *arithmetic* circuits;

$x \oplus y := x \vee y$ and $x \odot y := x \wedge y$        monotone *Boolean* circuits;

$x \oplus y := \min\{x, y\}$ and $x \odot y := x + y$      *tropical* (min, +) circuits;

$x \oplus y := \max\{x, y\}$ and $x \odot y := x + y$      tropical (max, +) circuits.

The adjective "monotone" indicates that there are no *negative* constants as inputs, in the case of arithmetic circuits, and there are no *negation* gates $\neg x = 1 - x$, in the case of Boolean circuits.

The Boolean semiring is both additively and multiplicatively idempotent, tropical semirings are additively but not multiplicatively idempotent, while arithmetic semiring is neither additively nor multiplicatively idempotent (see Table 1.1). The "multiplicative" neutral element $\mathbb{1}$ in both tropical semirings is 0 because $0 + x = x + 0 = x$. The "additive" neutral element $\mathbb{0}$ in (min, +) and (max, +) semirings is, respectively, $+\infty$ and $-\infty$, but we will *not* assume that these infinite "numbers" are present.

**Polynomials Produced by Circuits**

Every circuit $\Phi(x_1, \ldots, x_n)$ over a semiring $(R, \oplus, \odot)$ not only computes some function $f : R^n \to R$ but also *produces* (purely syntactically) an $n$-variate polynomial over this semiring in a natural way. Namely, at each source node $u$ holding a constant $c \in R$, the constant polynomial $P_u = c$ is produced, and at a source node $u$ holding a variable $x_i$, the polynomial $P_u = x_i$ is produced. At an "addition" gate $u = v \oplus w$, the "sum" $P_u = P_v \oplus P_w$ of the polynomials $P_v$ and $P_w$ produced at its inputs is produced. Finally, at a "multiplication" gate $u = v \odot w$, the polynomial $P_u$ obtained from $P_v \odot P_w$ by the distributivity of $\odot$ over $\oplus$ is produced; that is, we "multiply" ($\odot$) every term of $P_v$ with every term of $P_w$, and take the "sum" ($\oplus$) of the obtained terms. The polynomial produced by the entire circuit $\Phi$ is the polynomial

$$P(x) = \sum_{a \in A} c_a \prod_{i=1}^{n} x_i^{a_i}$$

produced at its output gate. Here, $A \subseteq \mathbb{N}^n$ is some set of *exponent vectors*, $x_i^k$ stands for the $k$-times "multiplication" $x_i \odot x_i \odot \cdots \odot x_i$, and $x_i^0 = 1$ (the multiplicative neutral element). Since by our assumption, the semiring contains the multiplicative neutral element $1$, coefficients $c_a$ are semiring elements[4]. The polynomial $P$ is *multilinear* if its set $A$ of exponent vectors consists of only 0-1 vectors (in each monomial, the power of every variable is at most one).

Two $n$-variate polynomials over a semiring $(R, \oplus, \odot)$ are *equivalent* (or represent the same function) if they take the same values on all inputs $a \in R^n$. A circuit $\Phi$ over a semiring $(R, \oplus, \odot)$ *computes* a given polynomial function $f : R^n \to R$ over this semiring if the polynomial $P$ produced by $\Phi$ is equivalent to $f$, that is, if $P(a) = f(a)$ holds for all $a \in R^n$.

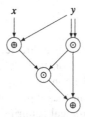

For example, the circuit depicted above produces the (generic) polynomial $P(x, y) = (x \oplus y) \odot y^2 \oplus y^2 = xy^2 \oplus y^3 \oplus y^2$. Over the tropical semiring $(\mathbb{R}_+, \max, +)$, the produced polynomial has the form $P(x, y) = \max\{\max\{x, y\} + 2y, 2y\} = \max\{x + 2y, 3y, 2y\}$ and is equivalent to the tropical $(\max, +)$ polynomial $f(x, y) = \max\{x + 2y, 3y\}$. Over the tropical semiring $(\mathbb{R}_+, \min, +)$, the produced polynomial has the form $P(x, y) = \min\{x + 2y, 3y, 2y\}$ and is equivalent to the (also tropical) polynomial $f(x, y) = 2y$ consisting of just one tropical monomial $2y$.

## 1.3   Tropical Circuits

As we already mentioned above, and as already the title of this book indicates, our main focus will be on tropical circuits. Recall that in the *tropical* $(\min, +)$ semiring, the domain is $\mathbb{R}_+$ and the operations are $x \oplus y := \min\{x, y\}$ and $x \odot y := x + y$ (see Table 1.2 for the correspondences between arithmetic and tropical semirings). Similarly, the domain in the tropical $(\max, +)$ semiring is also $\mathbb{R}_+$, and the operations are $x \oplus y := \max\{x, y\}$ and $x \odot y := x + y$.

---

[4] Because then, by distributivity, $x \oplus x = (1 \odot x) \oplus (1 \odot x) = (1 \oplus 1)x$, where $1 \oplus 1$ is a semiring element. An example where this is not the case is the semiring $(R, +, \times)$, where $R \subset \mathbb{N}$ is the set of all even integers: then the coefficient "3" of $x + x + x$ is not a semiring element.

**Table 1.2** Correspondences between the arithmetic and tropical semirings. In both cases, $A \subseteq \mathbb{N}^n$ is the set of "exponent" vectors of the corresponding "monomials," and $c_a$ are "coefficients" of these monomials

|  | Arithmetic $(+, \times)$ | Tropical $(\min, +)$ |
|---|---|---|
| Addition | $x + y$ | $\min\{x, y\}$ |
| Multiplication | $x \times y$ | $x + y$ |
| Division | $x/y$ | $x - y$ |
| Subtraction | $x - y$ | Undefined |
| Harmonic sum | $\dfrac{1}{\frac{1}{x} + \frac{1}{y}} = \dfrac{xy}{x+y}$ | $x + y - \min\{x, y\} = \max\{x, y\}$ |
| Monomials | $X^a = \prod_{i=1}^n x_i^{a_i}; \, a \in \mathbb{N}^n$ | $\langle a, x \rangle = \sum_{i=1}^n a_i x_i$ |
| Polynomials | $f(x) = \displaystyle\sum_{a \in A} c_a X^a; \, c_a \in \mathbb{R}_+$ | $f(x) = \displaystyle\min_{a \in A}\{\langle a, x \rangle + c_a\}$ |

Thus, a tropical $(\min, +)$ circuit $\Phi$ is a directed acyclic graph with each source node holding either a variable $x_i$ or a constant $c \in \mathbb{R}_+$. Every other node, a gate, has indegree two and performs either the min or $+$ operation. Such a circuit is *constant-free* if it has no constants other than 0 as inputs. In tropical $(\max, +)$ circuits, we have max instead of min. In the tropical semirings, "powering" $x_i^{a_i} = x_i \odot x_i \odot \cdots \odot x_i$ ($a_i \in \mathbb{N}$ times) turns into multiplication by scalars $a_i x_i = x_i + x_i + \cdots + x_i$. So, a (generic) monomial $\prod_{i=1}^n x_i^{a_i}$ turns into the scalar product

$$\langle a, x \rangle = a_1 x_1 + \cdots + a_n x_n$$

of vectors $a$ and $x$, and the tropical $(\max, +)$ and $(\min, +)$ versions of an arithmetic polynomial $P(x) = \sum_{a \in A} c_a \prod_{i=1}^n x_i^{a_i}$ are the *tropical polynomials*:

$$f(x) = \max_{a \in A} \langle a, x \rangle + c_a \quad \text{or} \quad f(x) = \min_{a \in A} \langle a, x \rangle + c_a \,,$$

where $A \subseteq \mathbb{N}^n$ is the set of "exponent" vectors of the polynomial, $\langle a, x \rangle$ are "monomials" with vectors $a \in A$ being their "exponent" vectors, and real scalars $c_a \in \mathbb{R}_+$ are the "coefficients" of these "monomials." That is, tropical circuits solve optimization problem with linear objective functions. For example, in the tropical $(\min, +)$ semiring, the arithmetic polynomial $P = (x + y)^2 = x^2 + 2xy + y^2$ turns into the tropical polynomial $f = \min\{2x, x + y + 2, 2y\}$ which, in its turn, is equivalent to the polynomial $g = \min\{2x, 2y\}$.

Functions represented by tropical polynomials are piecewise linear convex or concave functions. Recall that a function $f : \mathbb{R}_+^n \to \mathbb{R}_+$ is *convex* if

$$f(\lambda x + (1 - \lambda)y) \leqslant \lambda f(x) + (1 - \lambda) f(y)$$

holds for any vectors $x, y \in \mathbb{R}_+^n$ and any constant $0 \leqslant \lambda \leqslant 1$ and is *concave* if the converse inequality "$\geqslant$" holds.

**Proposition 1.2** *Functions computed by tropical* (max, +) *circuits are convex, while functions computed by tropical* (min, +) *circuits are concave.*

**Proof** Let $\Phi$ be a (max, +) circuits. The function $f : \mathbb{R}_+^n \to \mathbb{R}_+$ computed by $\Phi$ is of the form $f(x) = \max_{a \in A} \langle a, x \rangle + c_a$ for some $A \subseteq \mathbb{N}^n$ and $c_a \in \mathbb{R}_+$. Let $z := \lambda x + (1 - \lambda)y$ for some $x, y \in \mathbb{R}_+^n$ and $0 \leqslant \lambda \leqslant 1$. Take a vector $a \in A$ for which the maximum $f(z) = \langle a, z \rangle + c_a = \lambda\langle a, x \rangle + (1 - \lambda)\langle a, y \rangle + c_a$ on the input weighting $z$ is achieved. Then $\lambda f(x) + (1 - \lambda)f(y) \geqslant [\lambda\langle a, x \rangle + \lambda c_a] + [(1 - \lambda)\langle a, y \rangle + (1 - \lambda)c_a] = \lambda\langle a, x \rangle + (1 - \lambda)\langle a, y \rangle + c_a = f(z)$.

The proof that functions computed by tropical (min, +) circuits are concave is the same with "$\geqslant$" replaced by "$\leqslant$" in the above reasoning. □

**Example 1.3** The maximum function $f(x) = \max\{x_1, x_2\}$ cannot be computed by a (min, +) circuit because $f : \mathbb{R}_+^2 \to \mathbb{R}_+$ is not concave. Say, for $x = (2, 4)$ and $y = (4, 2)$, we have $f(\frac{1}{2}x + \frac{1}{2}y) = f(3, 3) = 3 < 4 = 2+2 = \frac{1}{2}f(x) + \frac{1}{2}f(y)$. □

A tropical circuit $\Phi$ *approximates* a given optimization problem $f : \mathbb{R}_+^n \to \mathbb{R}_+$ within a *factor* $r \geqslant 1$ (or *$r$-approximates* $f$) if for every input weighting $x \in \mathbb{R}_+^n$, the output value $\Phi(x)$ of the circuit lies:

o   Between $f(x)$ and $r \cdot f(x)$, in the case when $f$ is a *minimization* problem.
o   Between $f(x)/r$ and $f(x)$, in the case when $f$ is a *maximization* problem.

The factor $r$ may depend on the length $n$ of the inputs $x$, but not on the inputs $x$ themselves (*one* factor for *all* inputs). In both cases, the *smaller* the factor $r$ is, the better is the approximation. In particular, factor $r = 1$ means that the problem is (exactly) *solved*. Thus, a tropical circuit $\Phi$ *solves* an optimization problem $f$ if the tropical polynomial $P$ produced by the circuit $\Phi$ satisfies $P(x) = f(x)$ for all input weightings $x \in \mathbb{R}_+^n$.

**Remark 1.4 (Why Only *Nonnegative* Weights?)** The reason to require tropical circuits to solve optimization problems on only nonnegative weights is twofold. First, efficient dynamic programming algorithms usually work well on nonnegative weightings but fail when negative weights are allowed. For example, the Bellman–Ford–Moore algorithm (see Example 1.7 below) actually computes the minimum weight of an *s-t walk* of length at most $n - 1$ (one and the same edge can be traversed several times in a walk). If the weights are nonnegative, and since every walk between two nodes contains a simple path between these nodes, the minimum will always be achieved on a simple path. But this may not be the case if we have negative weights.

The second reason to only allow nonnegative weights is that if tropical circuits are required to solve a given 0/1 optimization problem on arbitrary input weightings (positive and negative), then their power is the *same* as that of so-called Minkowski circuits (see Lemma 1.28 in Sect. 1.9). These are monotone arithmetic $(+, \times)$ circuits "ignoring" coefficients and, unlike for tropical circuits, their power is now relatively well understood. □

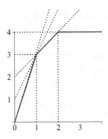

**Digression: Roots of Tropical Polynomials** A *root* of an $n$-variate arithmetic polynomial $f$ is a point $x_0 \in \mathbb{R}^n$ at which $f(x_0) = 0$. By Proposition 1.2, the function represented by a tropical (min, +) polynomial $f(x) = \min_{a \in A} \langle a, x \rangle + c_a$ is a concave piecewise linear function; functions represented by (max, +) are convex piecewise linear functions. A point $x_0 \in \mathbb{R}^n$ is a *root* of $f$ if $\langle a, x_0 \rangle + c_a = \langle b, x_0 \rangle + c_b = f(x_0)$ holds for some two distinct vectors $a \neq b \in A$. That is, roots in the tropical case are the points at which two "linear pieces" of $f$ intersect. For example, the univariate tropical (min, +) polynomial $f(x) = \min\{3x, 2x + 1, x + 2, 4\}$ has roots at $x = 1$ and $x = 2$.

The *individual degree* of an arithmetic polynomial $f$ is the smallest $d$ such that no single variable has an exponent larger than $d$ in $f$. Thus, the individual degree of a tropical polynomial $f(x) = \min_{a \in A} \langle a, x \rangle + c_a$ with $A \subseteq \mathbb{N}^n$ is $d = \max\{a_i : a \in A, i \in [n]\}$. In the classical (arithmetic) case, the well-known result, known as the DeMillo–Lipton–Schwartz–Zippel lemma [5, 27, 37], states that for every finite set $S \subseteq \mathbb{R}$ of size $|S| \geqslant d$, the maximal number of roots $x \in S^n$ of the (arithmetic) polynomial $\sum_{a \in A} c_a \prod_{i=1}^n x_i^{a_i}$ is $d|S|^{n-1}$. Grigoriev and Podolskii [12] have shown that, in the tropical case, the maximal possible number of roots is $|S|^n - (|S| - d)^n \approx nd|S|^{n-1}$. Non-roots of $f$ are weightings $x \in S^n$ that have *unique* feasible solutions $a \in A$: $\langle a, x \rangle + c_a = f(x)$ and $\langle a, x \rangle + c_a < \langle b, x \rangle + c_b$ for all $b \in A$, $b \neq a$. So, at least $(|S| - d)^n$ of all $|S|^n$ weightings $x \in S^n$ have unique solutions $a \in A$. This result is a generalization of the isolation lemma of Mulmuley, Vazirani, and Vazirani [20], which corresponds to the special case when $d = 1$, that is, when $A \subseteq \{0, 1\}^n$; a simpler proof of the isolation lemma itself was found by Ta-Shma [31].

## 1.4 Examples of Tropical Circuits

Recall that a combinatorial optimization problem is specified by a finite set of *ground elements* and a family of subsets of these elements, called *feasible solutions*. The problem itself is then, given an assignment of nonnegative real weights to the ground elements, to compute the minimum or the maximum weight of a feasible solution, the latter being the sum of weights of its elements.

Roughly speaking, the idea of DP is to break up the given optimization problem into smaller subproblems in a divide-and-conquer manner and solve these subproblems recursively. Optimal solutions of smaller instances are found and retained for use in solving larger instances (smaller instances are never solved again). Many classical DP algorithms are *pure* in that they only apply the basic operations (min, +) or (max, +) in their recursion equations. Note that these are the *only* operations used in the definitions of the optimization problems themselves.

Notable examples of pure DP algorithms for combinatorial optimization problems are the well-known Bellman–Ford–Moore shortest $s$-$t$ path algorithm [3, 10, 19], the Floyd–Warshall–Roy all-pairs shortest path algorithm [9, 23, 35], the Bellman–Held–Karp travelling salesman algorithm [4, 13], the Dreyfus–Levin–Wagner algorithm for the minimum weight Steiner tree problem [6, 18], and many others. Also the Viterbi (max, ×) DP algorithm [34]—which finds the most likely sequence of states resulting in a given sequence of observed events in a hidden Markov model—can be implemented as a pure (min, +) DP algorithm via the isomorphism $h : (0, 1] \to \mathbb{R}_+$ given by $h(x) = -\log x$.

Tropical (min, +) and (max, +) circuits constitute a rigorous mathematical model for pure DP algorithms. Namely, such an algorithm is just a special (recursively constructed) tropical circuit. In particular, lower bounds on the size of tropical circuit are also lower bounds on the minimum possible number of operations used by pure DP algorithms.

**Remark 1.5 (Why Only *Pure* DP?)** Why we restrict ourselves to "only" pure DP algorithms? Non-pure DP algorithms may use other arithmetic operations, as rounding, argmin, argmax, or, more crucially, conditional branchings (via if-then-else constraints). The presence of such operations makes the lower bounds problem for the corresponding circuit models no longer amenable for analysis using known mathematical tools. In particular, non-pure DP algorithms have the full power of general (non-monotone) Boolean circuits, for which even no super-linear lower bound is known so far. For example, Boolean negation $\neg x = 1 - x$ is a seemingly "innocent" conditional branching operation if $x = 0$ then 1 else 0. Let us stress that our goal is to prove *unconditional* lower bounds, not relying on unproven complexity assumptions like $\mathbf{P} \neq \mathbf{NP}$. In the context of this task, even proving lower bounds for (min, +, −) circuits (that is, tropical (min, +) circuits with subtraction gates allowed) remains a challenge (we shortly discuss the state of the art of this problem in Sect. 6.5). □

Let us recall some of the classical pure DP algorithms.

**Example 1.6 (Symmetric Polynomials)** The *elementary symmetric polynomial*

$$S_{n,k}(x_1, \ldots, x_n) = \sum_{S \subseteq [n]\ |S|=k} \prod_{i \in S} x_i$$

of degree $k$ (over any semiring) can be easily computed over *any* semiring $(R, \oplus, \odot)$ by using the following simple observation: every $r$-element subset $S \subseteq [m]$ is either an $r$-element subset of $[m-1]$ (if $m \notin S$) or is of the form $S = T \cup \{m\}$ for some $(r-1)$-element subset $T \subseteq [m-1]$. This gives us the following recursion equation:

$$S_{m,r}(x_1, \ldots, x_m) = S_{m-1,r}(x_1, \ldots, x_{m-1}) \oplus S_{m-1,r-1}(x_1, \ldots, x_{m-1}) \odot x_m .$$

Initial values are $S_{m,1}(x) = x_1 \oplus x_2 \oplus \cdots \oplus x_m$ for all $m = 1, \ldots, n$ and $S_{r,r}(x) = x_1 \odot x_2 \odot \cdots \odot x_r$ for all $r = 1, \ldots, k$; these values can be computed using $n+k$ gates. Thus, over any semiring (including the two tropical semirings), the polynomial $S_{n,k}$ can be computed using at most $\mathcal{O}(kn)$ gates. For example, over the tropical (max, +) semiring, the polynomial $S_{n,k}$ has the form $S_{n,k}(x_1, \ldots, x_n) = \max \left\{ \sum_{i \in S} x_i : S \subseteq [n], |S| = k \right\}$, and the recursion equation is

$$S_{m,r}(x_1, \dots, x_m) = \max \left\{ S_{m-1,r}(x_1, \dots, x_{m-1}), \ S_{m-1,r-1}(x_1, \dots, x_{m-1}) + x_m \right\}.$$

Initial values are $S_{m,1}(x) = \max\{x_1, x_2, \dots, x_m\}$ and $S_{r,r}(x) = x_1 + x_2 + \cdots + x_r$.
$\qquad\qquad\qquad\qquad\qquad\qquad\qquad\qquad\qquad\qquad\qquad\qquad\qquad\qquad\qquad\qquad$ □

**Example 1.7 (Single-Source Shortest Paths)** Let $K_n$ be the complete graph on $\{1, \dots, n\}$. The *shortest s-t path problem* for two fixed nodes $s \neq t \in [n]$ is, given an assignment of nonnegative weights $x_{i,j}$ to the edges $\{i, j\}$ of $K_n$, to compute the minimum weight of a path from node $s$ to node $t$, the latter being the sum of weights of its edges. The *single-source shortest path* problem SSSP($n$) is to (simultaneously) compute minimum weights of paths from a given source node $s$ to all remaining nodes $t \neq s$. The Bellman–Ford–Moore (min, +) DP algorithm [3, 10, 19] solves this problem using $\mathcal{O}(n^3)$ operations. The idea is very simple: if $P$ is a lightest path from $s$ to a node $j$ and if $\{i, j\}$ is the last edge of $P$, then the part of $P$ between $s$ and $i$ is a lightest path from $s$ to $i$.

To formalize this idea, consider the subproblems $W_k[i]$ for all nodes $i \neq s$ and integers $1 \leqslant k \leqslant n - 1$, where $W_k[i]$ is the minimum weight of a *walk* from $s$ to $i$ consisting of at most $k$ (not necessarily distinct) edges. Initial values are $W_1[i] = x_{s,i}$ for all $i \in [n] \setminus \{s\}$. The recursion equation is

$$W_k[j] = \min \left\{ W_{k-1}[j], \ \min_{i \neq j}\{W_{k-1}[i] + x_{i,j}\} \right\}.$$

The outputs are $W_{n-1}[t]$ for all nodes $t \neq s$. A fragment of the corresponding to this algorithm tropical (min, +) circuit is depicted in Fig. 1.1. Each min gate here has fan-in $n$ and is simulated by $n - 1$ fan-in-2 gates. So, the resulting (min, +) circuit has $\mathcal{O}(n^3)$ gates. The circuit actually computes the minimum weight of *walks* from the source node $s$ consisting of at most $n - 1$ (not necessarily distinct) edges. But every walk contains a (simple) path, and every simple path is also a walk. So, since the weights are nonnegative, the minimum will always be achieved on a (simple) path. Note that the same algorithm also works when the weight of a path is defined as the maximum weight of its edge (not as the sum of weights of these edges): it is then enough to replace $W_{k-1}[i] + x_{i,j}$ in the recursion equation by $\max\{W_{k-1}[i], x_{i,j}\}$. $\qquad\qquad\qquad\qquad\qquad\qquad\qquad\qquad\qquad\qquad\qquad\qquad\qquad\qquad$ □

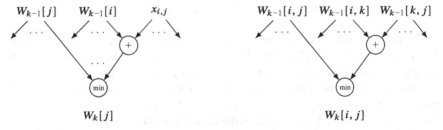

**Fig. 1.1** Fragments of tropical (min, +) circuits implementing the Bellman–Ford–Moore DP algorithm (left) and the Floyd–Warshall–Roy DP algorithm (right)

**Example 1.8 (All-Pairs Shortest Paths)** The *all-pairs shortest path* problem APSP($n$) on $K_n$ is, given an assignment of nonnegative weights $x_{i,j}$ to the edges $\{i, j\}$ of $K_n$, to simultaneously compute, for all pairs $\{i, j\}$ of nodes, the minimum weight of a path between $i$ and $j$. By taking $n$ copies of the (min, +) circuit arising from the Bellman–Ford–Moore (min, +) DP algorithm we just described, this problem can be solved by a (min, +) circuit of size $\mathcal{O}(n^4)$.

The Floyd–Warshall–Roy DP algorithm [9, 23, 35] uses another recursion equation which results in an $n$ times smaller (min, +) circuit with only $\mathcal{O}(n^3)$ gates. For $k \geqslant 1$ and nodes $i \neq j$, let $W_k[i, j]$ be the minimum weight of a path from $i$ to $j$ whose all inner nodes belong to the set $\{1, \ldots, k\}$. We have only two possibilities: either such a path does not go through the node $k$, or it does go through $k$ (from $i$ to $k$ and then from $k$ to $j$), both only using intermediate nodes in $\{1, \ldots, k - 1\}$. This simple observation gives us the recursion equation

$$W_k[i, j] = \min \{W_{k-1}[i, j], \ W_{k-1}[i, k] + W_{k-1}[k, j]\} \ .$$

As initial values we have $W_0[i, j] := x_{i,j}$. The outputs are $W_n[i, j]$ for all nodes $i \neq j$. A fragment of the corresponding to this algorithm tropical (min, +) circuit is depicted in Fig. 1.1. The same algorithm also works when the weight of a path is defined as the maximum weight of its edge (not as the sum of weights of these edges): it is then enough to replace $W_{k-1}[i, k] + W_{k-1}[k, j]$ in the recursion equation by $\max\{W_{k-1}[i, k], W_{k-1}[k, j]\}$.

We will latter show (Corollary 2.3 in Chap. 2) that the Floyd–Warshall–Roy pure DP algorithm is actually *optimal* in the class of all pure DP algorithms: every (min, +) circuit solving the all-pairs shortest path problem on $K_n$ must have $\Omega(n^3)$ gates.                                                                                                         □

**Example 1.9 (Traveling Salesman Problem)** We have to visit cities 1 to $n$. We start in city 1, run through the remaining $n-1$ cities in some order (without returning to any already visited city), and finally return to city 1. Inputs are nonnegative distances $x_{i,j}$ between cities $i$ and $j$. In the *traveling salesman problem* (TSP), the goal is to minimize the total travel length. A trivial algorithm for TSP checks all $(n - 1)! = \Omega((n/e)^n)$ permutations.

The Bellman–Held–Karp DP algorithm [4, 13] solves this problem by performing a still exponential, but much smaller number $\mathcal{O}(n^2 2^n)$ of operations. It takes as subproblems $W[S, i]$ = length of the shortest path that starts in city 1, then visits all cities in $S \setminus \{i\}$ in an arbitrary order, and finally stops in city $i$; here $S \subseteq \{2, \ldots, n\}$ and $i \in S$. Clearly, $W[\{i\}, i] = x_{1,i}$. The DP recursion is

$$W[S, i] = \min_{j \in S \setminus \{i\}} W[S \setminus \{i\}, j] + x_{j,i} \ .$$

The optimal travel length is given as the minimal value of $W[\{2, \ldots, n\}, i] + x_{i,1}$ over all $2 \leqslant i \leqslant n$. There are at most $n2^{n-1}$ subproblems $W[S, i]$, and each one uses $\mathcal{O}(n)$ gates to solve, *given* the solution of "smaller" subproblems.

Jerrum and Snir [14] have shown that $\Omega(n^2 2^n)$ (min, +) operations are also necessary to solve the TSP problem; we give a simple proof of a slightly weaker lower bound $2^{\Omega(n)}$ in Sect. 3.2 (Theorem 3.6). Thus, the Bellman–Held–Karp pure DP algorithm for the TSP problem is also optimal in the class of all pure DP algorithms.                                                                                     □

**Example 1.10 (MIS Problem on Trees)** The *maximum weight independent set* problem (*MIS problem*) on a graph $G = (V, E)$ is, given an assignment of weights $x_v \in \mathbb{R}_+$ to the nodes $v \in V$ (this time not to the edges) of $G$, to compute the maximum weight $\sum_{v \in I} x_v$ of an independent set $I \subseteq V$ in $G$; a set of nodes is *independent* if there are no edges between them. We will later show (Theorem 3.7 in Sect. 3.3) that the MIS problem on some $n$-vertex graphs $G$, including the $m \times m$ grid for $n = m^2$, requires (max, +) circuits of size $2^{\Omega(\sqrt{n})}$. In contrast, the MIS problem on any $n$-vertex *tree* $T = (V, E)$ can be solved by a (max, +) circuit of size $\mathcal{O}(n)$.

Fix an arbitrary vertex $r$ of $T$ as the root, and orient edges from the root to leaves. The subtree rooted at $v$, denoted by $T_v$, includes $v$ and all vertices reachable from $v$ under this orientation of edges. The *children* of $v$, denoted by $C_v$, are all those nodes $u$ entered by an edge from $v$. *Leaves* of the tree are nodes without children. Let $W[v]$ denote the maximum weight of an independent set of the subtree $T_v$. We want to compute $W[r]$. Let $W^+[v]$ denote the maximum weight of an independent set in $T_v$ that contains the node $v$, and let $W^-[v]$ denote the maximum weight of an independent set of $T_v$ that does not contain $v$. Initial values are $W^+[u] := x_u$ and $W^-[u] := 0$ for all leaves $u$ of $T$, and the recursions are

$$W^+[v] = x_v + \sum_{u \in C_v} W^-[u] \text{ and } W^-[v] = \sum_{u \in C_v} \max\left\{W^-[u], W^+[u]\right\}.$$

The output is $W[r] = \max\{W^-[r], W^+[r]\}$. Since $\sum_{v \in V} |C_v| \leqslant \sum_{v \in V} \deg(v) = 2|E| = 2(n-1)$, the number of (max, +) operations is $\mathcal{O}(n)$.                     □

## 1.5  Concentrate on Exponents: Minkowski Circuits

Our main goal in this book is to prove lower bounds on the size of tropical (min, +) and (max, +) circuits solving (or only approximating) optimization problems $f(x) = \min_{a \in A}\langle a, x\rangle$ or $f(x) = \max_{a \in A}\langle a, x\rangle$ on given sets $A \subseteq \mathbb{N}^n$ of feasible solutions. These are then also lower bounds on the number of operations that must be performed by *any* pure DP algorithm solving these problems.

It turns out that this goal can be achieved by solving an easier problem: prove that particular "similar" to $A$ sets $B \subseteq \mathbb{N}^n$ of vectors cannot be produced (purely syntactically) by small "Minkowski circuits." The aim of this section is to explain this general approach. For sets $A, B \subseteq \mathbb{R}^n$ of vectors, their *Minkowski sum* (or a *sumset*) is the set of vectors of the form

$$A + B = \{a + b \colon a \in A \text{ and } b \in B\},$$

where, as customary, $a + b = (a_1 + b_1, \ldots, a_n + b_n)$ is the componentwise sum of vectors $a$ and $b$. That is, we add *every* vector of $B$ to *every* vector of $A$.

Let $\Phi$ be a circuit over some semiring $(R, \oplus, \odot)$. As we already mentioned in Sect. 1.2, at each gate $u$, the circuit $\Phi$ *produces* (purely syntactically) some polynomial $P_u$ over $(R, \oplus, \odot)$. Of interest for us will be not as much the polynomials $P_u$ themselves but rather the sets of exponent vectors of these polynomials. These sets $X_u \subseteq \mathbb{N}^n$ of exponent vectors have a simple inductive construction, where $\vec{0}$ is the all-0 vector, and $\vec{e}_i \in \{0, 1\}^n$ has exactly one 1 in the $i$th position:

- $X_u = \{\vec{0}\}$ if $u = c \in R$ (input constant).
- $X_u = \{\vec{e}_i\}$ if $u = x_i$ (input variable).
- $X_u = X_v \cup X_w$ if $u = v \oplus w$ ("addition" gate).
- $X_u = X_v + X_w$ if $u = v \odot w$ ("multiplication" gate).

The set $A \subseteq \mathbb{N}^n$ produced by the entire circuit $\Phi$ is the set $X_o = A$ produced at the output gate $o$. The set of exponent vectors of a "sum" ($\oplus$) of two polynomials is the set-theoretical union of sets of exponent vectors of these polynomials. The exponent vector of a "product" of two monomials is the sum of their exponent vectors. Hence, the set of exponent vectors of a product ($\odot$) of two polynomials is the Minkowski sum of sets of exponent vectors of these polynomials. The set of vectors produced by the entire circuit is the set of exponent vectors of the polynomial produced at the output gate.

We thus have the following simple but important fact:

> If $A \subseteq \mathbb{N}^n$ is the set of exponent vectors produced by a circuit over some semiring, then the circuit computes some polynomial over that semiring whose set of exponent vectors is $A$.

It is clear that the same circuit $\Phi$ with "addition" ($\oplus$) and "multiplication" ($\odot$) gates can compute *different* functions $f : R^n \to R$ over different semirings. Say, the circuit $\Phi = x \odot y \oplus z$ computes $xy + z$ over the arithmetic $(+, \times)$ semiring but computes $\max\{x + y, z\}$ over the tropical $(\max, +)$ semiring. It is, however, important to note that (see Fig. 1.2):

> The set of exponent vectors *produced* by a circuit over any semiring is always the same—it only depends on the circuit itself, not on the underlying semiring.

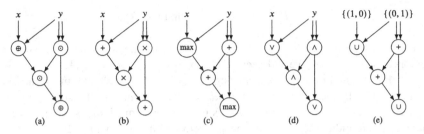

**Fig. 1.2** The same circuit (**a**) over three different semirings $(R, \oplus, \odot)$ and their Minkowski version (**e**). Here $\Downarrow$ stands for two parallel edges. The monotone arithmetic circuit (**b**) computes the polynomial $f(x, y) = (x+y)y^2 + y^2 = xy^2 + y^2 + y^3$. The tropical (max, +) circuit (**c**) solves the maximization problem $f(x, y) = \max\{x + 2y, 2y, 3y\} = \max\{x + 2y, 3y\}$. The monotone Boolean circuit (**e**) computes the monotone Boolean function $f(x, y) = (x \vee y) \wedge y \vee y = y$. All these circuits produce the same set of vectors $A = \{(1, 2), (0, 2), (0, 3)\}$, as done by the Minkowski version (**e**) of all these circuits

Note that the sets $X_u \subseteq \mathbb{N}^n$ of exponent vectors produced at gates $u$ of a circuit over any semiring are obtained by using only union ($\cup$) and Minkowski sum (+) operations. Thus, we arrive to a "semiring independent" model of circuits. By analogy with Boolean ($\vee, \wedge$) circuits, it is natural to call ($\cup, +$) circuits Minkowski circuits.

Formally, a *Minkowski* ($\cup, +$) *circuit* $\Phi$ is a directed acyclic graph with $n + 1$ input (indegree zero) nodes holding single-element sets $\{\vec{0}\}, \{\vec{e}_1\}, \ldots, \{\vec{e}_n\}$. Every other node, a *gate*, has indegree two and performs either the set-theoretic union ($\cup$) or the Minkowski sum (+) operation on its two inputs. At each gate of the circuit $\Phi$, some set of vectors is produced in an obvious way (by using these two operations $\cup$ and +). The set of vectors produced by the entire circuit is the set produced at the output gate. For a set $A \subseteq \mathbb{N}^n$ of vectors, let

$$L(A) := \text{min size of a Minkowski } (\cup, +) \text{ circuit producing the set } A.$$

The *Minkowski* ($\cup, +$) *version* of a circuit $\Phi$ over an arbitrary semiring $(R, \oplus, \odot)$ is obtained by replacements (where $c \in R$ is any "constant" input of $\Phi$): $c \mapsto \{\vec{0}\}$, $x_i \mapsto \{\vec{e}_i\}$, $u \oplus v \mapsto u \cup v$, and $u \odot v \mapsto u + v$. The following simple fact is the main motivation for considering Minkowski circuits: Minkowski version of a circuit $\Phi$ over any semiring produces the set of exponent vectors of the polynomial produced by $\Phi$.

**Lemma 1.11** *Let $\Phi$ be a circuit over some semiring. If $B \subseteq \mathbb{N}^n$ is the set of exponent vectors produced by $\Phi$, then* $\text{size}(\Phi) \geqslant L(B)$.

**Proof** There is a natural homomorphism from the semiring of $n$-variate polynomials over any semiring $(R, \oplus, \odot)$ to the semiring $(2^{\mathbb{N}^n}, \cup, +)$ of finite sets of vectors that maps every polynomial $f(x) = \sum_{b \in B_f} c_b \prod_{i=1}^{n} x_i^{b_i}$ to the set $B_f \subseteq \mathbb{N}^n$ of its exponent vectors. In particular, every single variable $x_i$ is mapped to $B_{x_i} = \{\vec{e}_i\}$, and every input constant $c \in R$ is mapped to $B_c = \{\vec{0}\}$. That this is indeed a homomorphism follows from easily verifiable equalities $B_{f \oplus h} = B_f \cup B_h$ and $B_{f \odot h} = B_f + B_h$, the latter sum being the Minkowski sum of the sets $B_f$ and $B_h$. Thus, the Minkowski version $\Phi'$ of the circuit $\Phi$ (of the same size a $\Phi$) produces the set $B_f$ of exponent vectors of the polynomial $f$ produced by $\Phi$. In particular, $size(\Phi) \geqslant L(B)$ holds.                                                                    □

Lemma 1.11 suggests the following way to prove lower bounds on the tropical $(\min, +)$ or $(\max, +)$ circuit complexity of the optimization problem $f(x) = \min_{a \in A} \langle a, x \rangle$ or $f(x) = \max_{a \in A} \langle a, x \rangle$ on a given set $A \subseteq \mathbb{N}^n$ of feasible solutions by proving lower bounds on the Minkowski circuit complexity $L(B)$ of particular set $B \subseteq \mathbb{N}^n$ of vectors.

We take an arbitrary (unknown to us) tropical circuit $\Phi$ solving the minimization problem $f_A(x) = \min_{a \in A} \langle a, x \rangle$ on a given set $A$ of feasible solutions, and let $B \subseteq \mathbb{N}^n$ be the set of "exponent" vectors *produced* by $\Phi$. By Lemma 1.11, we know that the circuit $\Phi$ must have at least $L(B)$ gates. Unfortunately, unlike for the (given) set $A$, we do not know what this set $B$ actually is! We only know that the optimization problems on both sets $A$ and $B$ of feasible solutions are the same (as functions), but the produced set $B$ may differ from the given set $A$ (see, for example, Fig. 1.3). And this is the main difficulty when trying to prove lower bounds on the size of tropical circuits: *different* sets of feasible solutions can define the *same* optimization problem $f : \mathbb{R}_+^n \to \mathbb{R}_+$. Thus, to show that any $(\min, +)$ circuit solving the minimization problem $f_A$ on a given set $A \subseteq \mathbb{N}^n$ of feasible solutions must have $\geqslant t$ gates, we will argue in two steps:

1. **Structure:** Show that the sets $B \subseteq \mathbb{N}^n$ of "exponent" vectors *produced* by $(\min, +)$ circuits *solving* $f_A$ have particular structural properties (depending on $A$).
2. **Syntactic lower bound:** Show that $L(B) \geqslant t$ holds for *any* set $B \subseteq \mathbb{N}^n$ of vectors with these properties.

In the next three sections, Sects. 1.6 to 1.8, we will establish the structural properties of sets $B \subseteq \mathbb{N}^n$ produced by arithmetic, Boolean, and tropical circuits (Step 1). General methods to prove lower bounds on $L(B)$ (Step 2) will be presented later in Chaps. 2–5.

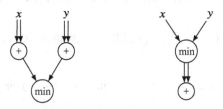

**Fig. 1.3** Two $(\min, +)$ circuits solving the minimization problem $f(x, y) = \min\{2x, 2y\}$ on the set $A = \{(2, 0), (0, 2)\}$ of feasible solutions. The first circuit produces the set $A$ itself, whereas the second saves one gate by producing a *different* set $B = \{(2, 0), (1, 1), (0, 2)\}$. Here (again) $\Downarrow$ stands for two parallel edges

## 1.6 Arithmetic Circuits Produce What They Compute

In general, the set of "exponent" vectors *produced* by a circuit does not need to coincide with the set of "exponent" vectors of the polynomial *computed* by that circuit (again, see Fig. 1.3 for a simple example). Arithmetic circuits make here a big exception: if such a circuit computes a polynomial, then it also produces that polynomial (as a formal expression). This is an easy consequence (Corollary 1.13) of the following fact.

**Lemma 1.12 (Alon and Tarsi [1])** *Let $f \in \mathbb{R}[x_1, \ldots, x_n]$ be a polynomial such that each variable appears in $f$ with degree at most $d$. Let $S \subseteq \mathbb{R}$ be a set of size $|S| \geqslant d + 1$. If $f(a) = 0$ for all $a \in S^n$, then $f(a) = 0$ for all $a \in \mathbb{R}^n$.*

***Proof*** We apply induction on $n$. For $n = 1$, the lemma is simply the assertion that a nonzero univariate polynomial of degree $d$ can have at most $d$ distinct roots. Assuming the lemma holds for $n - 1$, we prove it for $n$, $(n \geqslant 2)$. Given a polynomial $f(x) = f(x_1, \ldots, x_n)$ and a set $S$ satisfying the hypotheses of the lemma, let us write $f$ as a polynomial in $x_n$, that is, $f(x) = \sum_{i=0}^{d} f_i \cdot x_n^i$, where each $f_i$ is some polynomial on the first $n - 1$ variables, and the degree of each variable in $f_i$ is also at most $d$. Since $f(a) = 0$ for all $a \in S^n$, for each point $b \in S^{n-1}$, the univariate polynomial $f(b, x_n) = \sum_{i=0}^{d} f_i(b) \cdot x_n^i$ vanishes on all $x_n \in S$ and is thus a zero polynomial. Thus, $f_i(b) = 0$ for all $i = 1, \ldots, d$ and all points $b \in S^{n-1}$ and, by the induction hypothesis, each $f_i$ is a zero polynomial (vanishes on all points $b \in \mathbb{R}^{n-1}$), implying that $f$ is also a zero polynomial. $\square$

**Corollary 1.13** *If two arithmetic n-variate polynomials with positive coefficients take the same values on all inputs $x \in \mathbb{N}^n$, then they coincide as formal expressions.*

That is, not only the sets of exponent vectors of these polynomials but even the corresponding coefficients are the same.

***Proof*** Let $f$ and $g$ be two $n$-variate arithmetic polynomials with positive coefficients. Suppose that they are distinct, that is, do not coincide as formal expressions. Then, since the coefficients are positive, $h = f - g$ is a nonzero polynomial[5] of some (possibly large, but finite) degree $d$. When applied with any set $S \subseteq \mathbb{N}$ of size $|S| \geqslant d + 1$, Lemma 1.12 yields $h(a) \neq 0$ and, hence, also $f(a) \neq g(a)$ for some $a \in S^n$. $\square$

For a polynomial $f(x) = \sum_{a \in A} c_a \prod_{i=1}^{n} x_i^{a_i}$ with positive coefficients $c_a > 0$, let

$\texttt{Arith}(f) :=$ min size of a monotone arithmetic $(+, \times)$ circuit computing $f$.

Say that two (arithmetic) polynomials with positive coefficients are *similar* if they have the same monomials (with apparently different coefficients).

---

[5] This does not necessarily hold if also negative coefficients may be present. For example, as formal expressions, the polynomials $f$ and $g = x_1 - x_1 + f$ are distinct, but $f - g$ is a zero polynomial.

**Lemma 1.14** *For every finite set $A \subseteq \mathbb{N}^n$ of vectors, we have*

$$L(A) = \min \left\{ \mathtt{Arith}(f) \colon f(x) \text{ is similar to } p(x) = \sum_{a \in A} \prod_{i=1}^n x_i^{a_i} \right\}. \quad (1.2)$$

Thus, Minkowski circuits are, in fact, monotone arithmetic circuits "ignoring" coefficients. In particular, for every arithmetic polynomial $f(x) = \sum_{a \in A} c_a \prod_{i=1}^n x_i^{a_i}$ with positive coefficients $c_a > 0$, we have a lower bound $\mathtt{Arith}(f) \geqslant L(A)$.

*Proof* The inequality ($\leqslant$) in Eq. (1.2) follows directly from Corollary 1.13 (arithmetic circuits produce what they compute) and from Lemma 1.11 applied to the arithmetic semiring $(+, \times)$. To show the converse inequality ($\geqslant$) in Eq. (1.2), let $\Phi$ be a Minkowski $(\cup, +)$ circuit of size $L(A)$ producing the set $A$, and consider the arithmetic $(+, \times)$ version $\Phi'$ of $\Phi$ obtained as follows: replace the input $\{\vec{0}\}$ by constant 1, each input $\{\vec{e}_i\}$ by the variable $x_i$, every union gate $u \cup v$ by the (arithmetic) addition gate $u + v$, and every Minkowski sum gate $u + v$ by the (arithmetic) multiplication gate $u \times v$. The polynomial produced by the arithmetic circuit $\Phi'$ is then of the form $f(x) = \sum_{a \in A} c_a \prod_{i=1}^n x_i^{a_i}$ with some positive integer coefficients $c_a \in \mathbb{N}$ (indicating how often the same monomial appears in the produced polynomial). Thus, $L(A) = \text{size}(\Phi) = \text{size}(\Phi') \geqslant \mathtt{Arith}(f)$. Since the polynomial $f$ is similar to $p(x) = \sum_{a \in A} \prod_{i=1}^n x_i^{a_i}$, we are done. $\quad\square$

## 1.7 What Do Boolean Circuits Produce?

As shown by Corollary 1.13, every monotone arithmetic polynomial function has a *unique* representation as a polynomial. In contrast, over the Boolean $(\vee, \wedge)$ semiring, there is a much greater freedom in choosing representations for polynomial functions, that is, for Boolean functions. A Boolean function $f \colon \{0, 1\}^n \to \{0, 1\}$ is *monotone* if $b \leqslant a$ and $f(b) = 1$ imply $f(a) = 1$.

The *upward closure* of a set $A \subseteq \mathbb{N}^n$ is the set $A^\uparrow$ of all vectors $b \in \mathbb{N}^n$ such that $b \geqslant a$ for at least one vector $a \in A$. A set $A \subseteq \mathbb{N}^n$ of vectors is an *antichain* if $a, b \in A$ and $a \leqslant b$ implies $a = b$ (no two distinct vectors are comparable under $\leqslant$).

Let $\mathrm{Sup}(A) := \{\sup(a) \colon a \in A\} \subseteq 2^{[n]}$ denote the family of supports $\sup(a) := \{i \colon a_i \neq 0\}$ of vectors $a \in A$. In particular, $\mathrm{Sup}(A) \subseteq \mathrm{Sup}(B)$ means that for every vector $a \in A$, the set $B$ contains a vector with the same set of nonzero positions as $a$.

**Lemma 1.15 (Folklore)** *Let $A \subseteq \{0, 1\}^n$ be an antichain. A monotone Boolean circuit $\phi$ computes the Boolean function $f_A(x) = \bigvee_{a \in A} \bigwedge_{i \in \sup(a)} x_i$ iff the set $B \subseteq \mathbb{N}^n$ of exponent vectors produced by $\phi$ satisfies $\mathrm{Sup}(A) \subseteq \mathrm{Sup}(B)$ and $B \subseteq A^\uparrow$.*

**Proof** The Boolean function computed by the circuit $\phi$ is of the form $f_B(x) = \bigvee_{b \in B} \bigwedge_{i \in \sup(b)} x_i$. The "if" direction follows directly from the simple observation that $f_A(x) = 1$ iff $\sup(x) \supseteq \sup(a)$ for some $a \in A$. Hence, $\text{Sup}(A) \subseteq \text{Sup}(B)$ implies $f_A(x) \leqslant f_B(x)$, and $B \subseteq A^{\uparrow}$ implies $f_B(x) \leqslant f_A(x)$.

To show the "only if" direction, assume that the circuit $\phi$ computes $f_A$. If $b \notin A^{\uparrow}$ held for some vector $b \in B$, then on the input $x \in \{0, 1\}^n$ with $x_i = 1$ iff $i \in \sup(b)$, we would have $f_A(x) = 0$ while $f_B(x) = 1$. To show the inclusion $\text{Sup}(A) \subseteq \text{Sup}(B)$, suppose for a contradiction that there is a vector $a \in A$ such that $\sup(b) \neq \sup(a)$ holds for all vectors $b \in B$. Since $B \subseteq A^{\uparrow}$ and since $A$ is an antichain, $\sup(b) \subset \sup(a)$ (proper inclusion) cannot hold, for otherwise, the vector $a \in A$ would contain another vector of $A$. So, we have $\sup(b) \setminus \sup(a) \neq \emptyset$ for all vectors $b \in B$. But then $f_B(a) = 0$ while $f_A(a) = 1$, a contradiction. $\qquad \square$

## 1.8   What Do Tropical Circuits Produce?

Recall that the minimization (respectively, maximization) problem $f_A : \mathbb{R}_+^n \to \mathbb{R}_+$ on a set $A \subseteq \mathbb{N}^n$ of feasible solutions is, for every input weighting $x \in \mathbb{R}_+^n$, to compute the minimum (respectively, maximum) weight $\langle a, x \rangle = a_1 x_1 + \cdots + a_n x_n$ of a feasible solution $a \in A$.

Even if optimization problems on sets $A \subseteq \mathbb{N}^n$ and $B \subseteq \mathbb{N}^n$ of feasible solutions are equivalent ($f_B(x) = f_A(x)$ holds for all input weightings $x \in \mathbb{R}_+^n$), the structure of these sets may be very different. A simple example is given in Fig. 1.3. A more serious example will be given in Sect. 3.2 (see Theorem 3.6): there are explicit sets $A \subseteq \{0, 1\}^n$ and $B \subseteq \mathbb{N}^n$ (where $A$ is the set of feasible solutions of the shortest $s$-$t$ path problem) such that minimization problems $f_A$ and $f_B$ are equivalent, both can be solved by a (min, +) circuit of size $\mathcal{O}(n^{3/2})$, but $\text{L}(A)/\text{L}(B) = 2^{\Omega(\sqrt{n})}$ holds. This also shows that it can be exponentially harder to *produce* a set $A$ of feasible solutions by a (min, +) circuit than to *solve* the minimization problem on this set.

Still, the structure of sets $A$ and $B$ defining equivalent optimization problems $f_A$ and $f_B$ cannot be arbitrary: the sets $A$ and $B$ are then related via their convex hulls. This was shown by Jerrum and Snir [14] using a version of Farkas' lemma.

As customary, the product of a vector $a = (a_1, \ldots, a_n)$ with a scalar $\lambda \in \mathbb{R}$ is the vector $\lambda a = (\lambda a_1, \ldots, \lambda a_n)$, and the product of a set $A$ of vectors with $\lambda$ is the set of vectors $\lambda \cdot A = \{\lambda a : a \in A\}$. Recall that a vector $c \in \mathbb{R}^n$ is a *convex combination* of vectors[6] $\vec{b}_1, \ldots, \vec{b}_m$ in $\mathbb{R}^n$ if there are positive real scalars $\lambda_1, \ldots, \lambda_m > 0$ such that

$$\lambda_1 + \cdots + \lambda_m = 1 \quad \text{and} \quad c = \lambda_1 \vec{b}_1 + \cdots + \lambda_m \vec{b}_m .$$

---

[6] To keep the notation as simple as possible, we use the arrow notation $\vec{b}_i$ for vectors only when they are indexed.

It is not difficult to see the following *weighted average property*: for every convex combination $c = \lambda_1 \vec{b}_1 + \cdots + \lambda_m \vec{b}_m$ of some vectors in $B$, and for every vector $x \in \mathbb{R}^n$, we have

$$\min_{b \in B} \langle b, x \rangle \leqslant \langle c, x \rangle \leqslant \max_{b \in B} \langle b, x \rangle . \tag{1.3}$$

Indeed, to show the first inequality, let $1 \leqslant i_0 \leqslant m$ be such that $\langle \vec{b}_{i_0}, x \rangle = \min_i \langle \vec{b}_i, x \rangle$. Then $\langle c, x \rangle \geqslant \sum_{i=1}^{m} \lambda_i \langle \vec{b}_{i_0}, x \rangle = \langle \vec{b}_{i_0}, x \rangle \left( \sum_{i=1}^{m} \lambda_i \right) = \langle \vec{b}_{i_0}, x \rangle \geqslant \min_{b \in B} \langle b, x \rangle$. The proof of the second inequality in (1.3) is similar.

To relate the sets $A \subseteq \mathbb{N}^n$ and $B \subseteq \mathbb{N}^n$ of feasible solutions of equivalent discrete optimization problems, we will use the following form of Farkas' lemma [8] due to Fan [7, Theorem 4] (see also Schrijver [26, Corollary 7.1h]).

**Lemma 1.16 (Farkas [8], Fan [7])** *Let $u, \vec{u}_1, \ldots, \vec{u}_m \in \mathbb{R}^n$, and $\alpha, \alpha_1, \ldots, \alpha_m \in \mathbb{R}$. The following two assertions are equivalent:*

1. *$\forall y \in \mathbb{R}^n$ inequalities $\langle \vec{u}_1, y \rangle \geqslant \alpha_1, \ldots, \langle \vec{u}_m, y \rangle \geqslant \alpha_m$ imply $\langle u, y \rangle \geqslant \alpha$.*
2. *$\exists \lambda_1, \ldots, \lambda_m \in \mathbb{R}_+$ such that $u = \sum_i \lambda_i \vec{u}_i$ and $\alpha \leqslant \sum_i \lambda_i \alpha_i$.*

Nontrivial here is only the implication (1) $\Rightarrow$ (2). In the context of tropical circuits, the following consequence of Lemma 1.16 is important.

**Lemma 1.17 (Jerrum and Snir [14])** *Let $\vec{a}_0, \vec{a}_1, \ldots, \vec{a}_m \in \mathbb{R}^n$ and $c_0, c_1, \ldots, c_m \in \mathbb{R}$. The following two assertions are equivalent:*

1. *$\forall x \in \mathbb{R}_+^n : \langle \vec{a}_0, x \rangle + c_0 \geqslant \min_{1 \leqslant i \leqslant m} \langle \vec{a}_i, x \rangle + c_i$.*
2. *$\exists \lambda_1, \ldots, \lambda_m \in \mathbb{R}_+ : \sum_{i=1}^{m} \lambda_i = 1$ and $\vec{a}_0 \geqslant \sum_{i=1}^{m} \lambda_i \vec{a}_i$ and $c_0 \geqslant \sum_{i=1}^{m} \lambda_i c_i$.*

**Proof** The implication (2) $\Rightarrow$ (1) follows directly from the aforementioned property (1.3) of convex combinations. To show the converse implication (1) $\Rightarrow$ (2), observe that the assertion (1) is equivalent to the assertion that, for every $z \in \mathbb{R}$, the system of inequalities

$$\langle \vec{a}_i, x \rangle + c_i \geqslant z \qquad\qquad i = 1, \ldots, m,$$
$$\langle \vec{e}_j, x \rangle \geqslant 0 \qquad\qquad j = 1, \ldots, n,$$

implies the inequality $\langle \vec{a}_0, x \rangle + c_0 \geqslant z$. We use the inequalities $\langle \vec{e}_j, x \rangle \geqslant 0$ to ensure that we only consider vectors $x$ in $\mathbb{R}_+^n$ (with no negative entries). By taking

$$y := (x, z) ,$$
$$\vec{u}_i := (\vec{a}_i, -1) \text{ and } \alpha_i := -c_i \qquad\qquad i = 0, 1, \ldots, m,$$
$$\vec{u}_{m+j} := (\vec{e}_j, 0) \qquad\qquad j = 1, \ldots, n ,$$

each inequality $\langle \vec{a}_i, x \rangle + c_i \geqslant z$ turns into the inequality $\langle (\vec{a}_i, -1), (x, z) \rangle \geqslant -c_i$ and, hence, into the inequality $\langle \vec{u}_i, y \rangle \geqslant \alpha_i$. Thus, the above assertion turns into the assertion that for every vector $y$ in $\mathbb{R}^{n+1}$, the system of inequalities $\langle \vec{u}_i, y \rangle \geq \alpha_i$ for $i = 1, \ldots, m + n$ implies the inequality $\langle \vec{u}_0, y \rangle \geq \alpha_0$. Then, by Lemma 1.16, there exist $\lambda_1, \ldots, \lambda_{m+n} \in \mathbb{R}_+$ such that

$$(\vec{a}_0, -1) = \sum_{i=1}^{m} \lambda_i (\vec{a}_i, -1) + \sum_{j=1}^{n} \lambda_{m+j} (\vec{e}_j, 0) \quad \text{and} \quad -c_0 \leq \sum_{i=1}^{m} \lambda_i (-c_i).$$

This yields $\lambda_1 + \cdots + \lambda_m = 1, \vec{a}_0 \geq \sum_{i=1}^{m} \lambda_i \vec{a}_i$, and $c_0 \geq \sum_{i=1}^{m} \lambda_i c_i$.           $\square$

Say that a vector $a \in \mathbb{R}^n$ *contains* a vector $b \in \mathbb{R}^n$ if $a \geq b$ holds, that is, if $a_i \geq b_i$ holds for all $i = 1, \ldots, n$. For sets $A, B \subseteq \mathbb{R}^n$ of vectors, we say that:

- $B$ *lies above* $A$ if every vector of $B$ contains at least one vector of $A$.
- $B$ *lies below* $A$ if every vector of $B$ is contained in at least one vector of $A$.

Note that the fact that $B$ lies above $A$ does *not* imply that $A$ lies below $B$ and vice versa. The difference is in the order of quantifiers: the former means $\forall b \in B \,\exists a \in A : b \geq a$, while the latter means $\forall a \in A \,\exists b \in B : b \geq a$.

The main message of Lemma 1.17 is accumulated in the following lemma. Associate with a vector $x \in \mathbb{R}^n$ (an input weighting) its extension

$$\tilde{x} := (x, 1) \in \mathbb{R}^{n+1}.$$

For a set $U$ of real vectors, let $\text{Conv}(U)$ denote the set of all convex combinations of vectors in $U$, that is, the convex hull of $U$.

**Lemma 1.18** *Let $U, V \subseteq \mathbb{R}^{n+1}$ be finite sets of vectors. Then*

(i) $\min_{u \in U} \langle u, \tilde{x} \rangle \geq \min_{v \in V} \langle v, \tilde{x} \rangle$ *holds for all $x \in \mathbb{R}_+^n$ iff $U$ lies above $\text{Conv}(V)$.*
(ii) $\max_{u \in U} \langle u, \tilde{x} \rangle \leq \max_{v \in V} \langle v, \tilde{x} \rangle$ *holds for all $x \in \mathbb{R}_+^n$ iff $U$ lies below $\text{Conv}(V)$.*

***Proof*** Since $\min_{u \in U} \langle u, \tilde{x} \rangle \geq \min_{v \in V} \langle v, \tilde{x} \rangle$ iff $\langle u, \tilde{x} \rangle \geq \min_{v \in V} \langle v, \tilde{x} \rangle$ holds for every vector $u \in U$, the first claim (1) follows directly from Lemma 1.17. The second claim (2) also follows from Lemma 1.17 by using the equality $-\max\{x, y\} = \min\{-x, -y\}$. Thus, $\langle u, \tilde{x} \rangle \leq \max_{v \in V} \langle v, \tilde{x} \rangle$ holds iff $\langle -u, \tilde{x} \rangle \geq \min_{v \in V} \langle -v, \tilde{x} \rangle$. So, by applying Lemma 1.17 to the sets $-U = \{-u : u \in U\}$ and $-V$, we obtain that $-U$ lies *above* $\text{Conv}(-V) = -\text{Conv}(V)$ which (due to $-u \geq -v$ iff $u \leq v$) is equivalent to $U$ lying *below* $\text{Conv}(V)$.           $\square$

The following consequence of Lemma 1.18 shows that we can safely remove "redundant" terms from tropical polynomials without changing their functional behavior.

**Corollary 1.19 (Redundant Terms)** *Let $U \subseteq \mathbb{N}^n \times \mathbb{R}_+$ be finite, and $W \subsetneq U$.*

(i) $\min_{u \in U} \langle u, \tilde{x} \rangle = \min_{u \in U \setminus W} \langle u, \tilde{x} \rangle$ *holds for all $x \in \mathbb{R}_+^n$ iff $W$ lies above* $\text{Conv}(U \setminus W)$.

(ii) $\max_{u \in U} \langle u, \tilde{x} \rangle = \max_{u \in U \setminus W} \langle u, \tilde{x} \rangle$ *holds for all $x \in \mathbb{R}_+^n$ iff $W$ lies below* $\text{Conv}(U \setminus W)$.

*Proof* To show claim (i), let $f(x) = \min_{u \in U} \langle u, \tilde{x} \rangle$ and $g(x) = \min_{u \in U \setminus W} \langle u, \tilde{x} \rangle$. The inequality $g(x) \geq f(x)$ is trivial because $U \setminus W \subseteq U$ ($g$ has fewer feasible solutions). On the other hand, Lemma 1.18(i) and our assumption that the set $W$ lies above $\text{Conv}(U \setminus W)$ imply that the minimum of $\langle u, x \rangle$ over all vectors $u \in U$ is always achieved on a vector $u \in U \setminus W$. This gives the converse inequality $g(x) \leq f(x)$. The proof of claim (ii) is the same by using Lemma 1.18(ii). □

**Example 1.20** Consider the following two univariate $(\min, +)$ polynomials $f(x) = \min\{2x, x+1, 2\}$ and $g(x) = \min\{2x, x+2, 2\}$. The corresponding sets of vectors are $U = \{(2,0), (1,1), (0,2)\}$ and $V = \{(2,0), (1,2), (0,2)\}$. Since $(1,2) \geq (1,1) = \frac{1}{2}(2,0) + \frac{1}{2}(0,2)$, Corollary 1.19 implies that both polynomials $f$ and $g$ are (functionally) equivalent to the polynomial $h(x) = \min\{2x, 2\}$. □

Lemma 1.18 also yields the following tight characterization of the properties of sets of vectors produced by tropical circuits approximating given optimization problems. Recall that a *minimization problem* $f : \mathbb{R}_+^n \to \mathbb{R}_+$ on a set $A \subseteq \mathbb{N}^n$ (of feasible solutions) is of the form $f_A(x) := \min_{a \in A} \langle a, x \rangle$; in the case of maximization problems, we take max instead of min.

**Corollary 1.21** *Let $A, B \subseteq \mathbb{N}^n$ be finite sets of vectors.*

(1) *The minimization problem on $B$ approximates the minimization problem on $A$ within a factor $r \geq 1$ iff $B$ lies above $\text{Conv}(A)$ and $r \cdot A$ lies above $\text{Conv}(B)$.*

(2) *The maximization problem on $B$ approximates the maximization problem on $A$ within a factor $r \geq 1$ iff $B$ lies below $\text{Conv}(A)$ and $\frac{1}{r} \cdot A$ lies below $\text{Conv}(B)$.*

*Proof* To show claim (1), observe that, when applied with $U = B$ and $V = r \cdot A$, Lemma 1.18(i) implies that $f_B(x) \geq f_A(x)$ holds for all $x \in \mathbb{R}_+^n$ iff $B$ lies above $\text{Conv}(A)$. When applied with $U = r \cdot A$ and $V = B$, Lemma 1.18(i) implies that $r \cdot f_A(x) \geq f_B(x)$ holds for all $x \in \mathbb{R}_+^n$ iff $r \cdot A$ lies above $\text{Conv}(B)$.

To show claim (2), observe that, when applied with $U = \frac{1}{r} \cdot A$ and $V = B$, Lemma 1.18(ii) implies that $\frac{1}{r} \cdot f_A(x) \leq f_B(x)$ holds for all $x \in \mathbb{R}_+^n$ iff $\frac{1}{r} \cdot A$ lies below $\text{Conv}(B)$. When applied with $U = B$ and $V = A$, Lemma 1.18(ii) implies that $f_B(x) \leq f_A(x)$ holds for all $x \in \mathbb{R}_+^n$ iff $B$ lies below $\text{Conv}(A)$. □

### 1.8.1 Structure in 0/1 Optimization

Corollary 1.21 gives us a tight characterization of sets of vectors produced by tropical circuits, but its proof (relying on Farkas' lemma) is not very informative.

In particular, the proof stays silent about the following natural question: if a set of vectors produced by a tropical circuit violates claimed structural properties, on *what* input weightings the circuit must then err?

As observed in [15], in the case of 0/1 optimization problems (when the set $A$ of feasible solutions consists of only 0-1 vectors), a less tight but more informative characterization can be proved using elementary arguments (avoiding any use of Farkas' lemma). Lemmas 1.22 and 1.24 below are a bit more general than those in [15], but the ground idea of their proofs is the same.

The *upward closure* of a set $A \subseteq \mathbb{N}^n$ of vectors is

$$A^{\uparrow} := \{b \in \mathbb{N}^n : b \geqslant a \text{ for some } a \in A\},$$

and the *downward closure* of $A$ is $A^{\downarrow} := \{b \in \mathbb{N}^n : b \leqslant a \text{ for some } a \in A\}$. Note that $B \subseteq A^{\uparrow}$ iff $B$ lies above $A$, and $B \subseteq A^{\downarrow}$ iff $B$ lies below $A$. Recall that a set $A \subseteq \mathbb{N}^n$ of vectors is an *antichain* if $a, b \in A$ and $a \leqslant b$ implies $a = b$ (no two distinct vectors are comparable under $\leqslant$).

**Lemma 1.22 (Minimization)** *Let* $f_A(x) = \min_{a \in A} \langle a, x \rangle$ *and* $f_B(x) = \min_{b \in B} \langle b, x \rangle$ *be* (min, +) *polynomials, where* $A \subseteq \{0, 1\}^n$ *is an antichain and* $B \subseteq \mathbb{N}^n$. *Let* $r \geqslant 1$.

(i) *If* $f_A(x) \leqslant f_B(x) \leqslant r \cdot f_A(x)$ *holds for all* $x \in \{0, 1, rn + 1\}^n$, *then* $B \subseteq A^{\uparrow}$, *and for every vector* $a \in A$ *there is a vector* $b \in B$ *such that* $\sup(b) = \sup(a)$ *and* $\langle a, b \rangle \leqslant r \cdot \langle a, a \rangle$.

(ii) *If* $f_A(x) = f_B(x)$ *holds for all* $x \in \{0, 1, n + 1\}^n$, *then* $A \subseteq B \subseteq A^{\uparrow}$.

(iii) *If* $A \subseteq B \subseteq A^{\uparrow}$, *then* $f_A(x) = f_B(x)$ *holds for all* $x \in \mathbb{R}_+^n$.

*Proof*  To show claim (i), suppose that $f_A(x) \leqslant f_B(x) \leqslant r \cdot f_A(x)$ holds for all $x \in \{0, 1, rn + 1\}^n$. To show the inclusion $B \subseteq A^{\uparrow}$, take an arbitrary vector $b \in B$, and consider the input weighting $x \in \{0, 1\}^n$ such that $x_i := 0$ for $i \in \sup(b)$, and $x_i := 1$ for $i \notin \sup(b)$. Take a vector $a \in A$ on which the minimum $f_A(x) = \langle a, x \rangle$ is achieved. Then $\langle a, x \rangle = f_A(x) \leqslant f_B(x) \leqslant \langle b, x \rangle = 0$. Thus, $\sup(a) \subseteq \sup(b)$. Since $b \in \mathbb{N}^n$ and $a$ is a 0-1 vector, this yields $b \geqslant a$, as desired.

Now take an arbitrary vector $a \in A$, and consider the weighting $x \in \{1, rn + 1\}^n$ with $x_i := 1$ for all $i \in \sup(a)$ and $x_i := rn + 1$ for all $i \notin \sup(a)$. Let $b \in B$ be a vector on which the minimum $f_B(x) = \langle b, x \rangle$ is achieved. Then $\langle b, x \rangle = f_B(x) \leqslant r \cdot f_A(x) \leqslant r \cdot \langle a, x \rangle = r \cdot \langle a, a \rangle \leqslant rn$. If $b_i \geqslant 1$ held for some $i \notin \sup(a)$, then we would have $\langle b, x \rangle \geqslant b_i x_i = b_i(rn + 1) > rn$, a contradiction. Thus, the inclusion $\sup(b) \subseteq \sup(a)$ holds. Since $B \subseteq A^{\uparrow}$, there is a vector $a' \in A$ such that $a' \leqslant b$. Hence, $\sup(a') \subseteq \sup(b) \subseteq \sup(a)$. Since both $a$ and $a'$ are 0-1 vectors, this yields $a' \leqslant a$ and, since the set $A$ is an antichain, we have $a' = a$, and the equality $\sup(b) = \sup(a)$ follows. Hence, also $\langle b, x \rangle = \langle b, a \rangle$ holds, and we obtain $\langle a, b \rangle = f_B(x) \leqslant r \cdot \langle a, a \rangle$.

To show claim (ii), suppose that $f_B(x) = f_A(x)$ holds for all $x \in \{0, 1, n + 1\}^n$. In the proof of the inclusion $B \subseteq A^{\uparrow}$, we only used weightings $x \in \{0, 1\}^n$. So, it remains to show the inclusion $A \subseteq B$. For this, take an arbitrary vector $a \in A$. By using weightings $x \in \{1, n + 1\}^n$, we have shown in (ii) (for $r = 1$) that there is a vector $b \in B$ such that $\sup(b) = \sup(a)$ and $\langle a, b \rangle \leqslant \langle a, a \rangle$. Since $a$ is a 0-1 vector and $b \in \mathbb{N}^n$, this yields $b = a$, that is, vector $a$ belongs to $B$, as desired.

To show claim (iii), suppose that the inclusions $A \subseteq B \subseteq A^{\uparrow}$ hold, and take an arbitrary input weighting $x \in \mathbb{R}_+^n$. Then $A \subseteq B$ gives $f_A(x) \geqslant f_B(x)$, and $B \subseteq A^{\uparrow}$ gives $f_A(x) \leqslant f_B(x)$.                                                                   □

**Remark 1.23**  The proof of Lemma 1.22 shows that if $\exists b \in B \; \forall a \in A : \; b \not\geqslant a$, then the inequality $f_A(x) \leqslant f_B(x)$ is violated on the weighting $x \in \{0, 1\}^n$ with $x_i = 1$ iff $i \notin \sup(b)$. If $\exists a \in A \; \forall b \in B : \; \sup(a) \neq \sup(b)$, then the inequality $f_B(x) \leqslant r \cdot f_A(x)$ is violated on the weighting $x \in \{1, rn + 1\}^n$ with $x_i = 1$ for all $i \in \sup(a)$ and $x_i = rn + 1$ for all $i \notin \sup(a)$.                                      □

**Lemma 1.24 (Maximization)**  Let $f_A(x) = \max_{a \in A} \langle a, x \rangle$ and $f_B(x) = \max_{b \in B} \langle b, x \rangle$ be $(\max, +)$ polynomials with $A \subseteq \{0, 1\}^n$ and $B \subseteq \mathbb{N}^n$. Let $r \geqslant 1$.

(i) If $\frac{1}{r} \cdot f_A(x) \leqslant f_B(x) \leqslant f_A(x)$ holds for all $x \in \{0, 1\}^n$, then $B \subseteq A^{\downarrow}$, and for every vector $a \in A$ there is a vector $b \in B$ such that $\langle a, b \rangle \geqslant \frac{1}{r} \cdot \langle a, a \rangle$.

(ii) If $A$ is an antichain and $f_A(x) = f_B(x)$ holds for all $x \in \{0, 1\}^n$, then $A \subseteq B \subseteq A^{\downarrow}$.

(iii) If $A \subseteq B \subseteq A^{\downarrow}$, then $f_A(x) = f_B(x)$ holds for all $x \in \mathbb{R}_+^n$.

**Proof**  To show claim (i), observe that every vector $b \in B$ must be a 0-1 vector. Indeed, if $b$ had a position $i$ with $b_i > 1$, then on the weighting $x = \vec{e_i}$ we would have $f_B(x) > 1$ while $f_A(x) \leqslant 1$ since vectors in $A$ are 0-1 vectors.

To show the inclusion $B \subseteq A^{\downarrow}$, take an arbitrary vector $b \in B$. Then $f_B(b) \geqslant \langle b, b \rangle$. Let $a \in A$ be a vector on which the maximum $f_A(b) = \langle a, b \rangle$ is achieved. From $f_B(b) \leqslant f_A(b)$, we have $\langle b, b \rangle \leqslant \langle a, b \rangle$. Since both vectors $a$ and $b$ are 0-1 vectors, this yields $b \leqslant a$, as desired.

Now take an arbitrary vector $a \in A$, and let $b \in B$ be a vector on which the maximum $f_B(a) = \langle a, b \rangle$ is achieved. Then $\langle a, b \rangle = f_B(a) \geqslant \frac{1}{r} \cdot f_A(a) \geqslant \frac{1}{r} \cdot \langle a, a \rangle$, as desired.

To show claim (ii), let the approximation factor be $r = 1$, and suppose that $A$ is an antichain. We have only to show that then the inclusion $A \subseteq B$ holds. So, take an arbitrary vector $a \in A$. By (i) for factor $r = 1$, $\langle a, b \rangle \geqslant \langle a, a \rangle$ holds for some vector $b \in B$. Since both vectors $a$ and $b$ are 0-1 vectors, $\langle a, b \rangle \geqslant \langle a, a \rangle$ is equivalent to $b \geqslant a$. Since $B \subseteq A^{\downarrow}$ (again, by (i)), there must be a vector $a' \in A$ such that $b \leqslant a'$. Thus, $a \leqslant b \leqslant a'$. Since $A$ is an antichain, this yields $a = a'$ and, hence, $a = b$, as desired.

To show claim (iii), suppose that the inclusions $A \subseteq B \subseteq A^{\downarrow}$ hold, and take an arbitrary input weighting $x \in \mathbb{R}_+^n$. Then $A \subseteq B$ gives $f_A(x) \leqslant f_B(x)$, and $B \subseteq A^{\downarrow}$ gives $f_A(x) \geqslant f_B(x)$.                                                        □

**Remark 1.25**  The proof of Lemma 1.24(i) shows that if $\exists b \in B \; \forall a \in A : \; b \not\leqslant a$, then the inequality $f_B(x) \leqslant f_A(x)$ is violated on the weighting $x := b$, and if

**Table 1.3** Properties of sets $B \subseteq \mathbb{N}^n$ of "exponent" vectors produced by circuits over arithmetic, tropical, and Boolean semirings when computing a given multilinear polynomial $f(x) = \sum_{a \in A} \prod_{i=1}^{n} x_i^{a_i}$ over the corresponding semiring, where $A \subseteq \{0, 1\}^n$ is an antichain

| Circuits | Properties of produced sets $B$ | Reference |
|---|---|---|
| Arithmetic $(+, \times)$ | $A = B$ | Corollary 1.13 |
| Tropical $(\max, +)$ | $A \subseteq B \subseteq A^{\downarrow}$ | Lemma 1.24 |
| Tropical $(\min, +)$ | $A \subseteq B \subseteq A^{\uparrow}$ | Lemma 1.22 |
| Boolean $(\vee, \wedge)$ | $\mathrm{Sup}(A) \subseteq \mathrm{Sup}(B)$ and $B \subseteq A^{\uparrow}$ | Lemma 1.15 |

$\exists a \in A \; \forall b \in B : \; \langle a, b \rangle < \frac{1}{r} \cdot \langle a, a \rangle$, then the inequality $\frac{1}{r} \cdot f_A(x) \leqslant f_B(x)$ is violated on the weighting $x := a$. $\qquad\qquad\qquad\qquad\qquad\qquad\qquad\qquad\qquad\qquad\qquad$ □

The well-known *zero–one principle* for sorting networks (see, for example, Knuth [17, Theorem Z]) states that if a sorting network can correctly sort all $2^n$ sequences of zeros and ones, then it also sorts arbitrary inputs. Lemma 1.24 gives a similar principle for tropical $(\max, +)$ circuits.

**Corollary 1.26 (Zero–One Principle for (max,+) Circuits)** *Let $A \subseteq \{0, 1\}^n$ be an antichain. If a $(\max, +)$ circuit $\Phi$ solves the problem $f(x) = \max_{a \in A} \langle a, x \rangle$ on all 0-1 input weightings $x \in \{0, 1\}^n$, then $\Phi$ also solves the problem $f$ on all nonnegative real weightings $x \in \mathbb{R}_+^n$.*

***Proof*** Let $g(x) = \max_{b \in B} \langle b, x \rangle + c_b$ be the $(\max, +)$ polynomial produced by the circuit $\Phi$. Since all constants $c_b$ are nonnegative, and since $g(\vec{0}) = f(\vec{0}) = 0$ must hold, the polynomial $g$ is of the form $g(x) = \max_{b \in B} \langle b, x \rangle$ (all $c_b = 0$). Since $g(x) = f(x)$ for all $x \in \{0, 1\}^n$, Lemma 1.24(ii) gives the inclusions $A \subseteq B \subseteq A^{\downarrow}$. So, $g(x) = f(x)$ holds for all $x \in \mathbb{R}_+^n$. $\qquad\qquad\qquad\qquad\qquad\qquad\qquad\qquad\qquad$ □

Table 1.3 summarizes the properties of sets of "exponent" vectors produced by circuits over arithmetic, tropical, and Boolean semirings.

## 1.9  Negative Weights

Recall that a tropical $(\min, +)$ or $(\max, +)$ circuit $\Phi$ *solves* a given optimization problem $f$ if $\Phi(x) = f(x)$ holds for all nonnegative weightings $x \in \mathbb{R}_+^n$; on other (negative) weightings, the circuit may output any values. For a finite set $A \subseteq \mathbb{N}^n$ of vectors, let

$$\mathrm{Min}(A) := \text{min size of a } (\min, +) \text{ circuit solving } f(x) = \min_{a \in A} \langle a, x \rangle;$$

$$\mathrm{Max}(A) := \text{min size of a } (\max, +) \text{ circuit solving } f(x) = \max_{a \in A} \langle a, x \rangle.$$

In the case when circuits are required to solve a given problem $f$ on all input weightings $x \in \mathbb{R}^n$ (positive and negative), the corresponding complexity measures are denoted by $\text{Min}^-(A)$ and $\text{Max}^-(A)$. Note that we always have $\text{Min}(A) \leqslant$ $\text{Min}^-(A)$ and $\text{Max}(A) \leqslant \text{Max}^-(A)$: if a tropical circuit solves an optimization problem on all input weightings, then it also solves that problem on all nonnegative weightings.

**Lemma 1.27** *Let $A \subseteq \mathbb{N}^n$ be a finite set of vectors, $\Phi$ be a (min, +) circuit, and $B \subseteq \mathbb{N}^n$ be the set of "exponent" vectors produced by $\Phi$. Then $\Phi$ solves the minimization problem $f(x) = \min_{a \in A} \langle a, x \rangle$ on all input weightings $x \in \mathbb{R}^n$ iff the set $B$ satisfies $\text{Conv}(B) = \text{Conv}(A)$. The same also holds for (max, +) circuits.*

**Proof** Note that if the inequality in the first assertion (1) of Lemma 1.17 holds for *all* weightings $x \in \mathbb{R}^n$ (not only for nonnegative ones), then the same argument (by omitting the inequalities $\langle \vec{e}_j, x \rangle \geqslant 0$ ensuring the nonnegativity of weights) yields the *equality* $\vec{a}_0 = \sum_i \lambda_i \vec{a}_i$ (instead of inequality $\vec{a}_0 \geqslant \sum_i \lambda_i \vec{a}_i$) in the second assertion (2). Thus, for all vectors $\vec{a}_0, \vec{a}_1, \ldots, \vec{a}_m \in \mathbb{R}^n$ and constants $c_0, c_1, \ldots, c_m \in \mathbb{R}$, the following two assertions are equivalent:

(1) $\forall x \in \mathbb{R}^n : \langle \vec{a}_0, x \rangle + c_0 \geqslant \min_{1 \leqslant i \leqslant m} \langle \vec{a}_i, x \rangle + c_i$.

(2) $\exists \lambda_1, \ldots, \lambda_m \in \mathbb{R}_+ : \sum_i \lambda_i = 1$ and $\vec{a}_0 = \sum_i \lambda_i \vec{a}_i$ and $c_0 \geqslant \sum_i \lambda_i c_i$.

Now, take a (min, +) circuit $\Phi$ of size $\text{Min}^-(A)$ solving the minimization problem $f(x) = \min_{a \in A} \langle a, x \rangle$ for all input weightings $x \in \mathbb{R}^n$. Thus, the (min, +) polynomial $g(x) = \min_{b \in B} \langle b, x \rangle + c_b$ produced by $\Phi$ satisfies $g(x) = f(x)$ for all $x \in \mathbb{R}^n$. By the equivalence of (1) and (2), this can hold iff the inclusions $A \subseteq \text{Conv}(B)$ and $B \subseteq \text{Conv}(A)$ hold. Since $\text{Conv}(\text{Conv}(A)) = \text{Conv}(A)$, this is equivalent to $\text{Conv}(A) = \text{Conv}(B)$.

The proof for (max, +) circuits solving the corresponding maximization problem is the same by using $\max\{x, y\} = -\min\{-x, -y\}$. $\qquad\qquad\square$

For a (finite) set $A \subseteq \mathbb{N}^n$ of vectors, let

$$\text{L}_*(A) := \min \{\text{L}(B) : B \subseteq \mathbb{N}^n \text{ and } \text{Conv}(B) = \text{Conv}(A)\}$$

be the minimum size of Minkowski $(\cup, +)$ circuit producing a set $B \subseteq \mathbb{N}^n$ whose convex hull coincides with that of the set $A$.

**Lemma 1.28** *For every $A \subseteq \mathbb{N}^n$, both $\text{Max}^-(A)$ and $\text{Min}^-(A)$ lie between $\text{L}_*(A)$ and $L(A)$. If $A \subseteq \{0, 1\}^n$, then $\text{Max}^-(A) = \text{Min}^-(A) = L(A)$.*

**Proof** The inequalities $\text{L}_*(A) \leqslant \text{Min}^-(A)$ and $\text{L}_*(A) \leqslant \text{Min}^-(A)$ follow directly from Lemmas 1.11 and 1.27 applied to semirings $(\mathbb{R}, \min, +)$ and $(\mathbb{R}, \max, +)$.

To show the inequality $\text{Min}^-(A) \leqslant L(A)$, suppose that the set $A$ is produced by a Minkowski $(\cup, +)$ circuit $\Phi$ of size $L(A)$, and let $\Phi'$ be the (min, +) version of $\Phi$ obtained by replacements: $\{\vec{0}\} \mapsto 0$, $\{\vec{e}_i\} \mapsto \{x_i\}$, $u \cup v \mapsto \min\{u, v\}$, and $u + v \mapsto u + v$, where the first "+" in the latter replacement is the Minkowski sum of sets, while the second "+" is the ordinary (arithmetic) sum of numbers. (In the (max, +) version of the circuit $\Phi$, we use the replacement $u \cup v \mapsto \max\{u, v\}$.) Note that the obtained (min, +) circuit $\Phi'$ is constant-free, that is, has no nonzero constants as inputs. It remains to show that the obtained $(\min, +)$ circuit $\Phi'$ produces the tropical polynomial $f_A(x) = \min_{a \in A} \langle a, x \rangle$ whose set of "exponent" vectors is the set $A$ produced by the Minkowski circuit $\Phi$. This follows from trivial equalities $f_{\{\vec{0}\}}(x) = \langle \vec{0}, x \rangle = 0$, $f_{\{\vec{e}_i\}}(x) = \langle \vec{e}_i, x \rangle = x_i$ and the (also trivial) equalities holding for any sets $U, V \subseteq \mathbb{N}^n$ of vectors:

$$f_{U \cup V}(x) = \min_{u \in U \cup V} \langle u, x \rangle = \min \left\{ \min_{u \in U} \langle u, x \rangle, \min_{u \in V} \langle u, x \rangle \right\} = \min\{f_U(x), f_V(x)\},$$

$$f_{U+V}(x) = \min_{u \in U + V} \langle u, x \rangle = \min_{u \in U} \langle u, x \rangle + \min_{u \in V} \langle u, x \rangle = f_U(x) + f_V(x).$$

Now suppose that $A \subseteq \{0, 1\}^n$. It is enough to show that then $\mathrm{L}_*(A) \geqslant \mathrm{L}(A)$ holds. For this, it is enough to show that every set $B \subseteq \mathbb{N}^n$ satisfying $\mathrm{Conv}(B) = \mathrm{Conv}(A)$ must coincide with the set $A$. Since $A$ consists of only 0-1 vectors and $B$ consists of integer vectors, $B \subseteq \mathrm{Conv}(A)$ yields the inclusion $B \subseteq A$ because $\mathrm{Conv}(A) \cap \mathbb{N}^n = A$ (the interior of the unit cube does not contain lattice points). Together with $A \subseteq \mathrm{Conv}(B)$, the desired equality $B = A$ follows.                $\square$

## 1.10 Eliminating Constant Inputs

A tropical polynomial $f(x) = \min_{a \in A} \langle a, x \rangle + c_a$ or $f(x) = \max_{a \in A} \langle a, x \rangle + c_a$ is *constant-free* if $c_a = 0$ holds for all $a \in A$. A tropical circuit is *constant-free* if it uses no nonzero constants as inputs. Combinatorial optimization problems (the main object considered in this book) are constant-free, that is, are specified by constant-free tropical polynomials.

For example, in the famous *MST problem* (the minimum weight spanning tree problem) on a given graph $G$, the goal is to compute the constant-free $(\min, +)$ polynomial $f(x) = \min_{a \in A} \langle a, x \rangle$, where $A$ is the set of characteristic 0-1 vectors of spanning trees of $G$ (viewed as sets of their edges). In the not less prominent *assignment problem*, $A$ is the set of characteristic 0-1 vectors of perfect matchings, etc.

So, a natural question is can nonzero constant inputs in tropical circuits help when computing constant-free tropical polynomials? Lemma 1.29 below gives a *negative* answer:

> When dealing with tropical circuits solving or approximating optimization problems defined by constant-free tropical polynomials, we can safely restrict ourselves to constant-free circuits.

The *constant-free version* of a tropical circuit is obtained by replacing all constant inputs with constant 0.

**Lemma 1.29 (Jukna and Seiwert [16])** *If a tropical circuit $\Phi$ $r$-approximates a constant-free optimization problem, then the constant-free version of $\Phi$ also approximates this problem within the same factor $r$.*

*Proof* We consider two cases depending on whether $f$ is a maximization or minimization problem, the former case being almost trivial.

*Case 1:* $f(x) = \max_{a \in A} \langle a, x \rangle$ and $\Phi$ is a (max, $+$) circuit approximating $f$ within a factor $r$. Let $g(x) = \max_{b \in B} \langle b, x \rangle + c_b$ be the (max, $+$) polynomial produced by the circuit $\Phi$. Since the circuit $\Phi$ approximates $f$ within a factor $r$, we know that $f(x)/r \leqslant g(x) \leqslant f(x)$ holds for all input weightings $x \in \mathbb{R}_+^n$. Taking $x := \vec{0}$, we obtain $g(\vec{0}) \leqslant f(\vec{0}) = 0$ and, hence, $c_b = 0$ for all $b \in B$ (because the "coefficients" $c_b \in \mathbb{R}_+$ of the polynomial $g$ are nonnegative). Thus, in the case of maximization, the constant-free version $\Phi^o$ of the circuit $\Phi$ even *produces* the same polynomial $g$ as the original circuit $\Phi$.

*Case 2:* $f(x) = \min_{a \in A} \langle a, x \rangle$ and $\Phi$ is a (min, $+$) circuit approximating $f$ within a factor $r$. Let $g(x) = \min_{b \in B} \langle b, x \rangle + c_b$ be the (min, $+$) polynomial produced by the circuit $\Phi$. The constant-free version $\Phi^o$ of $\Phi$ produces the constant-free version $g^o(x) = \min_{b \in B} \langle b, x \rangle$ of the polynomial $g$ (constant inputs of the circuit $\Phi$ can only affect the "coefficients" $c_b$). Since the circuit $\Phi$ approximates $f$ within a factor $r$, we know that $f(x) \leqslant g(x) \leqslant r \cdot f(x)$ holds for all $x \in \mathbb{R}_+^n$. We have to show that the constant-free version $g^o$ of $g$ also satisfies these inequalities. Since the constants $c_b$ are nonnegative, we clearly have $g^o(x) \leqslant g(x)$ and, hence, also $g^o(x) \leqslant r \cdot f(x)$ for all $x \in \mathbb{R}_+^n$.

So, it remains to show that $f(x) \leqslant g^o(x)$ holds for all $x \in \mathbb{R}_+^n$, as well. To show this, we will exploit the fact that, because of constant-freeness, $f(\lambda x) = \lambda \cdot f(x)$ and $g^o(\lambda x) = \lambda \cdot g^o(x)$ hold for every scalar $\lambda \in \mathbb{R}$. Assume for the sake of contradiction that $f(x_0) > g^o(x_0)$ holds for some input weighting $x_0 \in \mathbb{R}_+^n$. Then the difference $d = f(x_0) - g^o(x_0)$ is positive. We can assume that the constant $c := \max_{b \in B} c_b$ is also positive, for otherwise, there would be nothing to prove. Take the scalar $\lambda := 2c/d > 0$. Since $g^o(x_0) = f(x_0) - d$, we obtain $g(\lambda x_0) \leqslant g^o(\lambda x_0) + c = \lambda \cdot g^o(x_0) + c = \lambda[f(x_0) - d] + c = f(\lambda x_0) - \lambda d + c = f(\lambda x_0) - c$, which is strictly smaller than $f(\lambda x_0)$, a contradiction with $f(x) \leqslant g(x)$ for all $x \in \mathbb{R}_+^n$.   $\square$

## 1.11   From Tropical to Boolean and Arithmetic Circuits

Every finite set $A \subseteq \mathbb{N}^n$ *defines* the monotone Boolean function

$$\hat{f}_A(x) = \bigvee_{a \in A} \bigwedge_{i \in \sup(a)} x_i \,,$$

where, as before, $\sup(a) = \{i : a_i \neq 0\}$ is the *support* of vector $a$. Recall that the minimization problem on $A$ is $f_A(x) = \min_{a \in A} \langle a, x \rangle$. Let

$\text{Min}_r(A) :=$ min size of a (min, +) circuit approximating the minimization

problem on $A$ within the factor $r$;

$\text{Bool}(A) :=$ min size of a monotone Boolean $(\vee, \wedge)$ circuit computing $\hat{f}_A$.

In particular, $\text{Min}(A) = \text{Min}_1(A)$ (factor $r = 1$).

**Lemma 1.30 (Boolean Bound)** *For every antichain $A \subseteq \{0, 1\}^n$ and every $r \geqslant 1$, we have $\text{Min}_r(A) \geqslant \text{Bool}(A)$.*

In Sect. 4.1.1 we will show that the same lower bound holds even for (min, max, +) circuits, i.e., for (min, +) circuits with max gates (Lemma 4.1).

***Proof*** Let $\Phi$ be a (min, +) circuit of size $s = \text{Min}_r(A)$ approximating the minimization problem $f_A(x) = \min_{a \in A} \langle a, x \rangle$ on $A$ with the factor $r$, and let $B \subseteq \mathbb{N}^n$ be the set of "exponent" vectors produced by $\Phi$. By Lemma 1.29, we can assume that the circuit $\Phi$ is constant-free. Hence, the minimization problem solved by the circuit $\Phi$ is of the form $f_B(x) = \min_{b \in B} \langle b, x \rangle$. Replace min gates by OR gates and + gates by AND gates. That is, we replace the "addition" gates by "addition" gates and "multiplication" gates by "multiplication" gates of the corresponding semirings. Thus, the obtained Boolean $(\vee, \wedge)$ version $\phi$ of the circuit $\Phi$ produces the same set $B$ of "exponent" vectors. Hence, the circuit $\phi$ computes the decision version $\hat{f}_B(x) = \bigvee_{b \in B} \bigwedge_{i \in \text{sup}(b)} x_i$ of the minimization problem $f_B$ on the set $B$. Since the tropical circuit $\Phi$ approximates the problem $f_A$, Lemma 1.22 implies that the set $B$ satisfies $\text{Sup}(A) \subseteq \text{Sup}(B)$ and[7] $B \subseteq A^\uparrow$. Thus, by Lemma 1.15, the Boolean circuit $\phi$ computes the decision version $\hat{f}_A$ of the minimization problem $f_A$, as desired.                                        $\square$

By Lemma 1.30, lower bounds on the tropical (min, +) circuit complexity of minimization problems can be obtained by proving lower bounds on the monotone Boolean circuit complexity of the decision versions of these problems. Unfortunately, the task of proving nontrivial lower bounds on the size of even monotone *Boolean* circuits remains a rather nontrivial task and, by this reason, only several such bounds are known so far. Fortunately, in some cases, lower bounds for tropical circuits can be obtained by proving lower bounds for Minkowski circuits (a.k.a. monotone arithmetic circuits) and, as we will see in subsequent chapters, the later task is often much easier than that for Boolean circuits.

**Lemma 1.31 (Arithmetic Bounds)** *For every antichain $A \subseteq \{0, 1\}^n$,*

$$\text{Max}(A) = \min\left\{L(B) \colon A \subseteq B \subseteq A^\downarrow\right\}; \tag{1.4}$$

$$\text{Min}(A) = \min\left\{L(B) \colon A \subseteq B \subseteq A^\uparrow\right\}. \tag{1.5}$$

---

[7] Recall that $b \in A^\uparrow$ iff $b \geqslant a$ for some $a \in A$, and $b \in A^\downarrow$ iff $b \leqslant a$ for some $a \in A$.

**Proof** To show Eq. (1.4), set $\ell := \min\{L(B): A \subseteq B \subseteq A^{\downarrow}\}$, and let $f_A(x) = \max_{a \in A}\langle a, x \rangle$ be the maximization problem on $A$. To show the inequality $\ell \leqslant \text{Max}(A)$, let $\Phi$ be a (max, +) circuit of size $t = \text{Max}(A)$ solving the problem $f_A$. By Lemma 1.29, we can assume that the circuit $\Phi$ is constant-free. Thus, the circuit $\Phi$ solves the maximization problem $f_B(x) = \max_{b \in B}\langle b, x \rangle$ on the set $B \subseteq \mathbb{N}^n$ of "exponent" vectors produced by $\Phi$. Since the circuit $\Phi$ solves the problem $f_A$, the problem $f_B$ is equivalent to $f_A$, that is, $f_A(x) = f_B(x)$ holds for all $x \in \mathbb{R}_+^n$. Then, by Lemma 1.24, the set $B$ satisfies $A \subseteq B \subseteq A^{\downarrow}$. By Lemma 1.11, we have $L(B) \leqslant \text{size}(\Phi) = t$, and the inequality $\ell \leqslant L(B) \leqslant \text{Max}(A)$ follows.

To show the converse inequality $\text{Max}(A) \leqslant \ell$, take a set $B \subseteq \mathbb{N}^n$ such that $A \subseteq B \subseteq A^{\downarrow}$ and $L(B) = \ell$. By Lemma 1.28, we have $\text{Max}(B) \leqslant L(B)$. On the other hand, since $A \subseteq B \subseteq A^{\downarrow}$, and since the input weights are nonnegative, the maximization problem on $B$ is equivalent to that on $A$. This yields $\text{Max}(A) = \text{Max}(B) \leqslant L(B) = \ell$, as desired.

The proof of Eq. (1.5) for (min, +) circuits is the same by using Lemma 1.22 instead of Lemma 1.24.                                                                           □

By Lemma 1.31, in order to prove a lower bound $\text{Max}(A) \geqslant t$ on the minimum size of a (max, +) circuit solving the maximization problem on a given set $A$ of feasible solutions, it is enough to show that $L(B) \geqslant t$ holds for *all* sets $B \subseteq \mathbb{N}^n$ satisfying $A \subseteq B \subseteq A^{\downarrow}$. We are now going to show that it is also enough to show that $L(B) \geqslant t$ holds for a specific subset $B \subseteq A$ of $A$, the so-called higher envelope of $A$; in case of (min, +) circuits, this holds for the "lower envelope" of $A$.

The *degree* of a vector $a \in \mathbb{N}^n$ is the sum $a_1 + \cdots + a_n$ of its entries. A set of vectors is *homogeneous* of degree $m$ if all its vectors have the same degree $m$. The *lower envelope* $\lfloor A \rfloor$ of a set $A$ is the set of all vectors in $A$ of minimum degree, and the *upper envelope* $\lceil A \rceil$ is the set of all vectors in $A$ of maximum degree. Note that both these sets are homogeneous. Also, if $A$ itself is homogeneous, then $\lfloor A \rfloor = \lceil A \rceil = A$.

**Lemma 1.32 (Jerrum and Snir [14, Theorem 2.4])** *For every finite set $B \subseteq \mathbb{N}^n$,*
$$L(B) \geqslant \max\{L(\lfloor B \rfloor), L(\lceil B \rceil)\}.$$

**Proof** From any $(\cup, +)$ circuit producing the set $B$, we can obtain $(\cup, +)$ circuits for $\lfloor B \rfloor$ and for $\lceil B \rceil$ by appropriately discarding some of the edges entering union $(\cup)$ gates, where *discarding* the edge $(w, v)$ entering a union gate $v = u \cup w$ means that we delete that edge, delete the $\cup$ operation labeling the gate $v$, and contract the other edge $(u, v)$.

This follows from simple properties of envelopes. First, the envelope of a Minkowski sum of sets is the Minkowski sum of their envelopes: $\lfloor A + B \rfloor = \lfloor A \rfloor + \lfloor B \rfloor$ and $\lceil A + B \rceil = \lceil A \rceil + \lceil B \rceil$ hold. This holds because the degree is additive: the degree of a sum of two vectors is the sum of their degrees. Second, for the union, we have $\lfloor A \cup B \rfloor = \lfloor A \rfloor$ if the minimum degree of a vector in $A$ is smaller than the minimum degree of a vector in $B$, $\lfloor A \cup B \rfloor = \lfloor B \rfloor$ if the minimum degree of a vector in $B$ is smaller than the minimum degree of a vector in $A$, and

$\lfloor A \cup B \rfloor = \lfloor A \rfloor \cup \lfloor B \rfloor$ otherwise, similarly for higher envelopes by looking at maximum (instead of minimum) degrees.                                                                    □

The following simple lemma shows that, when dealing with measures $\mathrm{Min}(A)$, $\mathrm{Max}(A)$, and $\mathrm{Bool}(A)$, one can safely assume that the set $A \subseteq \mathbb{N}^n$ of feasible solutions forms an antichain. The *lower antichain*[8] $A' \subseteq A$ of a set $A \subseteq \mathbb{N}^n$ of vectors consists of all vectors $a \in A$ such that $b \not\leqslant a$ for all $b \in A \setminus \{a\}$. That is, the set $A'$ is obtained by removing from $A$ all vectors containing some other vector of $A$. Similarly, the *higher antichain* $A' \subseteq A$ consists of all vectors $a \in A$ such that $b \not\geqslant a$ for all $b \in A \setminus \{a\}$. That is, the set $A'$ is obtained by removing from $A$ all vectors contained in some other vector of $A$.

**Lemma 1.33 (Antichain Lemma)** *Let $A \subseteq \mathbb{N}^n$ be a finite set of vectors.*

o *If $A' \subseteq A$ is the lower antichain of $A$, then $\mathrm{Min}(A') = \mathrm{Min}(A)$, $\mathrm{Bool}(A') = \mathrm{Bool}(A)$ and $\lfloor A' \rfloor = \lfloor A \rfloor$.*
o *If $A' \subseteq A$ is the higher antichain of $A$, then $\mathrm{Max}(A') = \mathrm{Max}(A)$ and $\lceil A' \rceil = \lceil A \rceil$.*

**Proof** Let $A' \subseteq A$ be the lower antichain of $A$. Since input weightings $x \in \mathbb{R}_+^n$ are nonnegative, $b \leqslant a$ implies $\langle b, x \rangle \leqslant \langle a, x \rangle$. Hence, the minimization problems on $A$ and on $A'$ are (functionally) equivalent; so, $\mathrm{Min}(A') = \mathrm{Min}(A)$. Since $b \leqslant a$ implies $\sup(b) \subseteq \sup(a)$, monotone Boolean functions defined by $A'$ and by $A$ are the same (as functions), meaning that $\mathrm{Bool}(A') = \mathrm{Bool}(A)$. Since no vector of smallest degree is removed from $A$ when forming $A'$, we also have $\lfloor A' \rfloor = \lfloor A \rfloor$. The case of higher antichains is similar.                                                         □

**Lemma 1.34 (Envelope Bounds)** *For every set $A \subseteq \{0, 1\}^n$, we have*

$$L(\lceil A \rceil) \leqslant \mathrm{Max}(A) \leqslant L(A) \quad \text{and} \quad L(\lfloor A \rfloor) \leqslant \mathrm{Min}(A) \leqslant L(A).$$

*If $A$ is also homogeneous, then $\mathrm{Max}(A) = \mathrm{Min}(A) = L(A)$.*

**Proof** Upper bounds $\mathrm{Max}(A) \leqslant L(A)$ and $\mathrm{Min}(A) \leqslant L(A)$ are given by Lemma 1.28 and are included just for completeness. So, it remains to prove the lower bounds $\mathrm{Max}(A) \geqslant L(\lceil A \rceil)$ and $\mathrm{Min}(A) \geqslant L(\lfloor A \rfloor)$. By the antichain lemma (Lemma 1.33), we can assume that $A$ is an antichain. Since $A$ consists of only 0-1 vectors and is an antichain, Lemma 1.31 gives the equality $\mathrm{Max}(A) = L(B)$ for some set $B \subseteq \mathbb{N}^n$ of vectors such that $A \subseteq B \subseteq A^\downarrow$. These two inclusions imply

---

[8] Lower *antichains* $A'$ of sets $A \subseteq \mathbb{N}^n$ should not be mixed with their lower *envelopes* $\lfloor A \rfloor$: we always have $\lfloor A \rfloor \subseteq A'$, but the equality does not need to hold. For example, if $A = \{(1, 0, 0), (0, 1, 0), (0, 0, 2), (1, 0, 2)\}$, then $\lfloor A \rfloor = \{(1, 0, 0), (0, 1, 0)\}$ but $A' = \{(1, 0, 0), (0, 1, 0), (0, 0, 2)\}$, similarly for higher antichains and higher envelopes.

that $\lceil B \rceil = \lceil A \rceil$. Thus, $\text{Max}(A) = \text{L}(B) \geqslant \text{L}(\lceil B \rceil) = \text{L}(\lceil A \rceil)$, where the inequality follows from Lemma 1.32. The proof of $\text{Min}(A) \geqslant \text{L}(\lfloor A \rfloor)$ is the same. If $A$ is homogeneous, then $\lfloor A \rfloor = \lceil A \rceil = A$ and, hence, the proved inequalities turn into equalities.                                                                                   □

**Remark 1.35** The condition in Lemma 1.34 that the set $A$ consists of only 0-1 vectors is important because there *exist* homogeneous sets $A \subseteq \mathbb{N}^n$ for which the gap $\text{L}(A)/\text{Min}(A)$ is even exponential. To see this, let $d = n/2$ and $B = \{\vec{b}_1, \ldots, \vec{b}_n\}$ be the set of $n$ vectors $\vec{b}_i = (0, \ldots, 0, d, 0, \ldots, 0)$, with $d$ in the $i$th position. Let also $E_{n,d} = \{a \in \mathbb{N}^n : a_1 + a_2 + \cdots + a_n = d\}$ be the set of all $|E_{n,d}| = \binom{n+d-1}{d} \geqslant (n/d)^d = 2^{n/2}$ vectors of degree $d$, and consider the family $\mathcal{A}$ of all $|\mathcal{A}| = 2^{|E_{n,d}|-|B|} = 2^{2^{\Omega(n)}}$ sets $A$ of vectors such that $B \subseteq A \subseteq E_{n,d}$. By Proposition 1.1, there are at most $s^{\mathcal{O}(s)}$ Minkowski $(\cup, +)$ circuits of size $\leqslant s$. So, in order to produce all sets $A \in \mathcal{A}$, we must have $s^{\mathcal{O}(s)} \geqslant |\mathcal{A}|$, and so $s \geqslant 2^{\Omega(n)}$. Thus, some sets $A \in \mathcal{A}$ (in fact, most of these sets) require Minkowski circuits of size $\text{L}(A) = 2^{\Omega(n)}$.

On the other hand, we have $\text{Min}(A) = \mathcal{O}(n)$ for every set $A \in \mathcal{A}$. Indeed, the minimization problem $f_B(x) = \min\{dx_1, dx_2, \ldots, dx_n\}$ on the set $B$ can be solved by a trivial $(\min, +)$ circuit of size $n + 2\log d$: first compute $y = \min\{x_1, \ldots, x_n\}$ using $n - 1$ gates, and then compute $f_B(x) = y + y + \cdots + y$ ($d$ times) using additional $2\log d$ gates. To show that this simple circuit also solves the minimization problem $f_A(x) = \min_{a \in A} \langle a, x \rangle$ on *any* set $A \in \mathcal{A}$, take an input weighting $x \in \mathbb{R}_+^n$. Then $f_B(x) \geqslant f_A(x)$ holds because $B \subseteq A$. To show $f_B(x) \leqslant f_A(x)$, take a vector $a \in A$ on which the minimum $f_A(x) = \langle a, x \rangle$ is achieved, and let $x_{i_0} = \min\{x_i : i \in S\}$ where $S = \sup(a)$. Then $f_A(x) = \langle a, x \rangle = \sum_{i \in S} a_i x_i \geqslant x_{i_0} \sum_{i \in S} a_i = dx_{i_0} \geqslant f_B(x)$.                                                                                   □

Finally, let us note that, in the case of Minkowski circuits, there is a more flexible analogue of Lemma 1.32 allowing one to concentrate on *any* envelope, not just on the lower or higher one. For sets $A_1, \ldots, A_m \subseteq \mathbb{N}^n$ of vectors, let $\text{L}(A_1, \ldots, A_m)$ denote the minimum size of a Minkowski $(\cup, +)$ circuit simultaneously producing all these sets (at some gates). The *$r$th envelope* of a set $A \subseteq \mathbb{N}^n$ of vectors is the set $A^{\langle r \rangle} \subseteq A$ of all vectors $a \in A$ of degree $a_1 + \cdots + a_n = r$. Hence, if $r$ and $d$ are the minimum and the maximum degrees of vectors in $A$, then $A^{\langle r \rangle} = \lfloor A \rfloor$, $A^{\langle d \rangle} = \lceil A \rceil$, and $A = A^{\langle r \rangle} \cup A^{\langle r+1 \rangle} \cup \cdots \cup A^{\langle d \rangle}$.

**Lemma 1.36 (Strassen [29])** *Let $A \subseteq \mathbb{N}^n$ be a finite set of vectors of maximum degree $d$. For every $r \leqslant d$, we have $\text{L}(A^{\langle 1 \rangle}, \ldots, A^{\langle r \rangle}) \leqslant (r+1)^2 \cdot \text{L}(A)$.*

In particular, for every $r \leqslant d$, we have a lower bound $\text{L}(A) \geqslant \text{L}(A^{\langle r \rangle})/(r+1)^2$.

*Proof* Let $\Phi$ be a $(\cup, +)$ circuit of size $s = \text{L}(A)$ producing $A$. We construct the desired circuit $\Phi'$ simultaneously producing all envelopes $A^{\langle 1 \rangle}, \ldots, A^{\langle r \rangle}$ as follows. For every gate $v$ in $\Phi$, we include $r + 1$ gates in $\Phi'$, which we denote $v_0, \ldots, v_r$. The goal is to achieve the situation where at the $i$th copy $v_i$ of a gate $v$ of $\Phi$, the $i$th envelope $X_v^{\langle i \rangle}$ of the set $X_v \subseteq \mathbb{N}^n$ of vectors produced at the gate $v$ in $\Phi$ is produced. This can be achieved by appropriately connecting the gates in the new circuit $\Phi'$. We construct $\Phi'$ inductively as follows.

If $v$ is an input gate holding a set $\{\vec{e}_i\}$, then we let $v_1 = v$, and for every $i \neq 1$, we let $v_i$ be an input gate holding the empty set $\emptyset$. If $v$ is an input gate holding a set $\{\vec{0}\}$, then we let $v_0 = v$, and for every $i \neq 0$, we let $v_i$ be an input gate holding the empty set $\emptyset$. Suppose now that $v$ is a gate of $\Phi$ entered from gates $u$ and $w$ and that we already have the desired copies $u_0, u_1, \ldots, u_r$ and $w_0, w_1, \ldots, w_r$ of the gates $u$ and $w$ in $\Phi'$. If $v$ is a union gate, then just take $v_i = u_i \cup w_i$ for all $i = 0, 1, \ldots, r$. If $v$ is a Minkowski sum gate, then let $v_i$ be the union of all Minkowski sums $u_l + w_{i-l}$ for $l = 0, 1, \ldots, i$ (Minkowski sum with the empty set is empty set). Induction implies that the circuit $\Phi'$ has the claimed functionality. Each gate of $\Phi$ is replaced by at most $\sum_{i=0}^{r}(i+1) \leqslant (r+1)^2$ gates in $\Phi'$. So, the new circuit $\Phi'$ has at most $(r+1)^2 s$ gates.                                                   □

**Remark 1.37** By Lemma 1.36, lower bounds on the Minkowski circuit complexity $L(A)$ of a set $A \subseteq \{0,1\}^n$ of vectors can be obtained by proving a lower bound on the Minkowski circuit complexity $L(A^{(r)})$ of any of its envelopes $A^{(r)}$. This no longer holds for tropical (min, +) circuits: here the gap $\mathrm{Min}(A^{(r)})/\mathrm{Min}(A)$ can be even exponential. Let, for example, $A \subseteq \{0,1\}^{\binom{n}{2}}$ be the set of characteristic 0-1 vectors of simple paths from vertex 1 to vertex $n$ in the complete graph $K_n$ on $\{1, \ldots, n\}$. Observe that the vectors of the $r$th envelope $A^{(r)}$ of $A$ for $r = n-1$ correspond to Hamiltonian paths from vertex 1 to vertex $n$, and we will show (Theorem 3.6) that $\mathrm{Min}(A^{(r)}) = 2^{\Omega(n)}$ holds. But, as we have already seen (Example 1.7), the Bellman–Ford–Moore (min, +) DP algorithm gives $\mathrm{Min}(A) = \mathcal{O}(n^3)$.                                             □

## 1.12  Coefficients Can Matter: Arithmetic $\neq$ Minkowski

For an $n$-variate polynomial $f$, let $A_f \subseteq \mathbb{N}^n$ denote the set of its exponent vectors. Thus, $f$ is of the form $f(x) = \sum_{a \in A_f} c_a \prod_{i=1}^{n} x_i^{a_i}$ for some coefficients $c_a > 0$; we call polynomials with positive coefficients *monotone polynomials*. By Lemma 1.14, we always have $\mathrm{Arith}(f) \geqslant L(A_f)$, where (as before) $\mathrm{Arith}(f)$ denotes the minimum size of a monotone arithmetic (+, ×) circuit computing $f$.

In fact, almost all known lower bounds on the monotone arithmetic (+, ×) circuit complexity of polynomials $f$, including [11, 14, 21, 24, 28, 32, 33], are lower bounds on the Minkowski circuit complexity $L(A_f)$ of their sets $A_f \subseteq \mathbb{N}^n$ of exponent vectors. Thus, each of these lower bounds for a polynomial $f$ holds also for *any* monotone polynomial with the same monomials as $f$. In particular, we can "wipe out" all coefficients from $f$ (replace them by constant 1), and the same lower bound holds for the obtained polynomial $g(x) = \sum_{a \in A_f} \prod_{i=1}^{n} x_i^{a_i}$. That is, the aforementioned lower bound arguments "ignore coefficients."

An argument used in a recent paper [36] by Yehudayoff is a notable exception. It shows that the gap $\mathrm{Arith}(f)/L(A_f)$ can be even exponential. The corresponding polynomial $f$ has $n^2$ variables $x_{i,j}$ arranged into an $n \times n$ matrix and is of the form

$$f(x) = 2^{-n} \sum_{a \in \{0,1\}^n} \prod_{i=1}^{n} \sum_{j=1}^{n} a_j x_{i,j} = 2^{-n} \sum_{a} \sum_{\varphi:[n] \to [n]} \prod_{i=1}^{n} a_{\varphi(i)} x_{i,\varphi(i)}$$

$$= 2^{-n} \sum_{\varphi} 2^{n-|\varphi([n])|} \prod_{i=1}^{n} x_{i,\varphi(i)} = \sum_{\varphi} c_\varphi \prod_{i=1}^{n} x_{i,\varphi(i)} \, ,$$

where the sum is taken over all $n^n$ mappings $\varphi : [n] \to [n]$, coefficients are[9] $c_\varphi = 2^{-|\varphi([n])|}$, and $\varphi([n]) = \{\varphi(i) : i \in [n]\}$ is the range of $\varphi$.

Note that the Minkowski circuit complexity of the set $A_f$ of exponent vectors of the polynomial $f$ is small, namely, is $\mathrm{L}(A_f) \leqslant n^2$. Indeed, if we wipe out all coefficients from the polynomial $f$ (replace them by constant 1), then we obtain the polynomial

$$g(x) = \sum_\varphi \prod_{i=1}^n x_{i,\varphi(i)} = \prod_{i=1}^n (x_{i,1} + x_{i,2} + \cdots + x_{i,n})$$

with the same set $A_f$ of exponent vectors as $f$. Hence, $\mathrm{L}(A_f) = \mathrm{L}(A_g) \leqslant \mathtt{Arith}(g) \leqslant n^2$. That is, the set of monomials of the polynomial $f$ has very simple structure. Still, the coefficients to these monomials already make the polynomial $f$ hard to compute by monotone arithmetic circuits.

**Theorem 1.38 (Yehudayoff [36])** $\mathtt{Arith}(f) = 2^{\Omega(n/\log n)}$.

*Proof* See Sect. 3.6.                                                         □

Together with Lemma 1.14, Theorem 1.38 shows that the monotone arithmetic circuit complexity of similar polynomials can be exponentially different; recall that two polynomials are *similar* if they have the same monomials with, apparently, different positive coefficients.

Actually, the fact that "coefficients can matter" even in *non-monotone* arithmetic circuits was (implicitly) shown by early results in algebraic circuit complexity. For example, already in 1974, Strassen [30] has shown the following lower bound for univariate polynomials. Recall that arithmetic (as well as tropical) circuits can use arbitrarily large constants as inputs for free; so, the largeness of coefficients $c_s$ in the following theorem is not a "problem:" the difficulty comes from the form of these coefficients.

**Theorem 1.39 (Strassen [30])** *Let $d \in \mathbb{N}$ be large enough. Every arithmetic $(+, -, \times, /)$ circuit computing the univariate polynomial $P_d(z) = \sum_{s=0}^d c_s z^s$ with coefficients $c_s = 2^{2^s}$ must have at least $\sqrt{d/(3\log d)}$ gates.*

Now, we can take $d = 2^n - 1$ and turn the polynomial $P_d(z)$ into an $n$-variate multilinear polynomial

$$f_n(x_1, \ldots, x_n) = \sum_{s=0}^{2^n-1} c_s \prod_{i=1}^n x_i^{\bar{s}_i},$$

---

[9] This form of coefficients comes from the fact that every $k$-element subset of $[n]$ is contained in $2^{n-k}$ subsets of $[n]$.

where $(\bar{s}_1, \ldots, \bar{s}_n) \in \{0, 1\}^n$ is the binary representation of the integer $s = \sum_{i=1}^{n} \bar{s}_i 2^{i-1}$. To transform a circuit $\Phi(x_1, \ldots, x_n)$ that computes the $n$-variate polynomial $f_n(x_1, \ldots, x_n)$ into a circuit $\Phi'(z)$ that computes the corresponding univariate polynomial $P_d(z)$, one can use a kind of the Kronecker substitution. Namely, it is enough to connect the inputs of the circuit $\Phi$ to the outputs of a circuit that computes the $n$-tuple of powers $z^{2^i}$, $i = 0, \ldots, n - 1$; such a circuit can be designed using $n - 1$ multiplication gates since $z^{2^{i+1}} = z^{2^i} \cdot z^{2^i}$. Then the circuit $\Phi(z^{2^0}, z^{2^1}, \ldots, z^{2^{n-1}})$ computes $P_d(z)$. Thus, Theorem 1.39 gives the following lower bound.

**Corollary 1.40** *Every arithmetic* $(+, -, \times, /)$ *circuit computing the polynomial* $f_n$ *must have* $2^{\Omega(n)}$ *gates. In particular,* $\texttt{Arith}(f_n) \geq 2^{\Omega(n)}$.

On the other hand, if we wipe out the coefficients $c_s$ from the polynomial $f_n$ (replace them by 1), then the resulting polynomial

$$g(x) = \sum_{s=0}^{2^n-1} \prod_{i=1}^{n} x_i^{\bar{s}_i} = 1 + \sum_{\emptyset \neq S \subseteq [n]} \prod_{i \in S} x_i = (1 + x_1)(1 + x_2) \cdots (1 + x_n)$$

can be computed by a trivial circuit of size at most $2n$. The set $A_g$ of exponent vectors of $g$ is the entire cube $\{0, 1\}^n = \{\vec{0}, \vec{e}_1\} + \{\vec{0}, \vec{e}_2\} + \cdots + \{\vec{0}, \vec{e}_n\}$. So, $L(A_g) \leq 2n$.

**Remark 1.41** Using the so-called representation theorem by Schnorr [25], Baur [2] gave a simple linear algebra proof of a general result which yields a full row of univariate polynomials $P(z) = \sum_{i=1}^{d} c_i z^i$ requiring arithmetic $(+, -, \times, /)$ circuits of size $\Omega(\sqrt{d/\log d})$. Such is, for example, the polynomial with coefficients $c_i = \sqrt{p_i}$, where $p_i$ is the $i$th prime.                                                                    □

**Remark 1.42 (VP $\neq$ VNP in the Monotone World)** Let us note that the goal of the aforementioned paper [36] was not to resolve the "can coefficients matter?" issue. The goal was rather to resolve a special (monotone) case of the algebraic version of the "P vs. NP" problem. An algebraic analogue of (Boolean) complexity class P is the class VP consisting of all polynomials computable by arithmetic circuits of size polynomial in the number of variables. The class VNP consists of all polynomials $f(x_1, \ldots, x_n)$ for which there is a polynomial $p(x, y) = p(x_1, \ldots, x_n, y_1, \ldots, y_m)$ in VP with $m = \text{poly}(n)$ additional variables such that $f(x) = \sum_{e \in \{0,1\}^m} p(x, e)$. In the monotone version of the class VNP, the defining polynomials $p(x, y)$ are monotone (have no negative coefficients). The Yehudayoff polynomial $f$ is in monotone VNP with the corresponding polynomial $p(x, y) = \prod_{i=1}^{n} \sum_{j=1}^{n} x_{i,j} y_j$ in VP (there are $n^2$ $x$-variables and $n$ $y$-variables). So, his lower bound (which we prove in Sect. 3.6) shows that VP $\neq$ VNP holds in the monotone world. Note that the separation "monotone VP $\neq$ monotone VNP" *cannot* be shown by "ignoring coefficients" arguments solely based on the structure of exponent vectors (that is, by considering the Minkowski circuit complexity) because, in the monotone world, the polynomials $f(x)$ and $p(x, 1, \ldots, 1)$ have the *same* sets of exponent vectors.                                                                    □

## 1.13  Complexity Gaps: An Overview

We summarize known bounds on the monotone Boolean, tropical, and monotone arithmetic circuit complexities of some polynomials in Table 1.4. Here, the *s-t path polynomial*, the *spanning tree polynomial*, and the *permanent polynomial* are, respectively, the multilinear polynomials

$$\mathrm{PATH}_n(x) = \sum_P \prod_{e \in P} x_e, \quad \mathrm{ST}_n(x) = \sum_T \prod_{e \in T} x_e, \quad \mathrm{PER}_n(x) = \sum_M \prod_{e \in M} x_e,$$

where the sums are, respectively, over all simple paths $P$ from vertex $s = 1$ to vertex $t = n$ in the complete undirected graph $K_n$ on the vertex-set $\{1, \ldots, n\}$, over all $n^{n-2}$ spanning trees $T$ of $K_n$, and over all $n!$ perfect matchings $M$ in $K_{n,n}$. The *squares polynomial* and the *no-isolated-vertex* polynomial are, respectively,

$$\mathrm{SQ}_n(x) = \sum_{\emptyset \neq S \subseteq [n]} \prod_{i,j \in S} x_{i,j} \quad \text{and} \quad \mathrm{ISOL}_n(x) = \sum_E \prod_{e \in E} x_e,$$

where in $\mathrm{ISOL}_n$, the summation is over all subsets $E$ of edges of $K_{n,n}$ with the property that for every vertex $v$ of $K_{n,n}$ there is an edge $e \in E$ with $v \in e$; that is, each $E$ is a subgraph of $K_{n,n}$ with no isolated vertices. All these polynomials are multilinear. The polynomials $\mathrm{ST}_n$ and $\mathrm{PER}_n$ are also homogeneous, while the other three are not. The corresponding tropical versions of these polynomials represent 0/1 optimization problems on their sets of exponent 0-1 vectors.

The following is a summary of known general relations between the powers of various kinds of circuits; all they are proved in this book. Let $A \subseteq \{0, 1\}^n$ be an antichain and $p(x) = \sum_{a \in A} c_a \prod_{i=1}^n x_i^{a_i}$ be any (multilinear) polynomial with the set $A$ of exponent vectors and some positive coefficients $c_a > 0$. Recall that $\mathrm{Bool}(A)$ denotes the minimum size of monotone Boolean circuit computing the Boolean function $f(x) = \bigvee_{a \in A} \bigwedge_{i \in \mathrm{sup}(a)} x_i$ defined by the set $A$. Then,

**Table 1.4** Some known bounds on the monotone Boolean $(\vee, \wedge)$, tropical $(\min, +)$ and $(\max, +)$, and monotone arithmetic $(+, \times)$ circuit complexities of some basic polynomials. In $(\min, \max, +)$ circuits, both min and max can be used as gates, and they must compute the $(\min, +)$ versions of the corresponding polynomials. The complexity of such a (minimization) version of $\mathrm{ISOL}_n$ in the case of $(\min, \max, +)$ circuits remains unclear

|  | $(\vee, \wedge)$ | $(\min, \max, +)$ | $(\min, +)$ | $(\max, +)$ | $(+, \times)$ | Ref. |
|---|---|---|---|---|---|---|
| $\mathrm{PATH}_n$ | $\mathcal{O}(n^3)$ | $\mathcal{O}(n^3)$ | $\mathcal{O}(n^3)$ | $2^{\Omega(n)}$ | $2^{\Omega(n)}$ | Theorem 3.6 |
| $\mathrm{ST}_n$ | $\mathcal{O}(n^3)$ | $\mathcal{O}(n^3)$ | $2^{\Omega(\sqrt{n})}$ | $2^{\Omega(\sqrt{n})}$ | $2^{\Omega(\sqrt{n})}$ | Theorems 3.16 and 6.5 |
| $\mathrm{PER}_n$ | $n^{\Omega(\log n)}$ [22] | $n^{\Omega(\log n)}$ | $2^{\Omega(n)}$ | $2^{\Omega(n)}$ | $2^{\Omega(n)}$ | Theorem 3.6, Corollary 4.3 |
| $\mathrm{SQ}_n$ | $\mathcal{O}(n^2)$ | $\mathcal{O}(n^2)$ | $\mathcal{O}(n^2)$ | $\mathcal{O}(n^2)$ | $n^{\Omega(\sqrt{n})}$ | Corollary 2.5 |
| $\mathrm{ISOL}_n$ | $\mathcal{O}(n^2)$ | ? | $2^{\Omega(n)}$ | $\mathcal{O}(n^2)$ | $2^{\Omega(n)}$ | Lemma 1.43 |

$$\text{Bool}(A) \overset{(1)}{\leqslant} \text{Min}(A) \overset{(2)}{\leqslant} \text{Min}^-(A) \overset{(3)}{=} \text{Max}^-(A) \overset{(4)}{=} \text{L}(A) \overset{(5)}{\leqslant} \text{Arith}(p).$$

If $A \subseteq \{0, 1\}^n$ is homogeneous, then

$$\text{Min}(A) \overset{(6)}{=} \text{Max}(A) \overset{(7)}{=} \text{L}(A).$$

Moreover, the gaps (1), (2), and (5) can be exponential.

Inequality (1) is a special case of Lemma 1.30 (when $r = 1$). Inequality (2) is trivial. Equalities (3) and (4) are given by Lemma 1.28. Inequality (5) is given by Lemma 1.14. Equalities (6) and (7) for homogeneous sets $A$ are given by Lemma 1.34.

That the gap (5) can be exponential is shown by Theorem 1.38. That the gap (2) can be such is shown by the squares polynomial $\text{SQ}_n$ (Corollary 2.5). To show that the gap (1) can also be exponential, consider the set $A \subseteq \{0, 1\}^{n \times n}$ of exponent vectors of the multilinear polynomial $\text{ISOL}_n$. Note that $A$ consists of all Boolean $n \times n$ matrices $a = (a_{i,j})$ such that each row and each column of $a$ has at least one 1.

**Lemma 1.43** We have $\text{Min}(A) = 2^{\Omega(n)}$ but $\text{Bool}(A) = \mathcal{O}(n^2)$.

***Proof*** The smallest number of 1s in a matrix $a \in A$ is $n$, and the matrices in $A$ with this number of 1s are permutation matrices. So, the lower envelope $\lfloor A \rfloor \subseteq A$ of $A$ consists of all permutation matrices, i.e., $\lfloor A \rfloor$ is the set of exponent vectors of the permanent polynomial $\text{PER}_n$. The lower bound $\text{L}(\lfloor A \rfloor) = 2^{\Omega(n)}$ was shown by Jerrum and Snir [14] (see also Theorem 3.6 in Sect. 3.2 for a simple proof). So, by Lemma 1.34, we have $\text{Min}(A) \geqslant \text{L}(\lfloor A \rfloor) = 2^{\Omega(n)}$. On the other hand, the Boolean function $f(x) = \bigvee_{a \in A} \bigwedge_{i \in \sup(a)} x_i$ defined by the set $A$ can be computed by the circuit $\phi(x) = \bigwedge_{i=1}^n \left( \bigvee_{j=1}^n x_{i,j} \right) \bigwedge_{j=1}^n \left( \bigvee_{i=1}^n x_{i,j} \right)$ of size $\mathcal{O}(n^2)$. $\quad\square$

A *formula* is a circuit with all its non-output gates of fanout 1. That is, the underlying graph is then a tree. In Sect. 5.5 (Corollary 5.11 and Remark 5.12) we will give an explicit optimization problem showing that:

Tropical *circuits* can be super-polynomially smaller than tropical *formulas*.

# References

1. Alon, N., Tarsi, M.: Colorings and orientations of graphs. Combinatorica **12**, 125–134 (1992)
2. Baur, W.: Simplified lower bounds for polynomials with algebraic coefficients. J. Complex. **13**, 38–41 (1997)
3. Bellman, R.: On a routing problem. Q. Appl. Math. **16**, 87–90 (1958)
4. Bellman, R.: Dynamic programming treatment of the travelling salesman problem. J. ACM **9**(1), 61–63 (1962)
5. DeMillo, R.A., Lipton, R.J.: A probabilistic remark on algebraic program testing. Inf. Process. Lett. **7**(4), 193–195 (1978)
6. Dreyfus, S., Wagner, R.: The Steiner problem in graphs. Networks **1**(3), 195–207 (1971)
7. Fan, K.: On systems of linear inequalities. In: Kuhn, H.W., Tucker, A.W. (eds.) Linear Inequalities and Related Systems. Ann. of Math. Stud., vol. 38, pp. 99–156. Princeton University Press, Princeton (1956)
8. Farkas, J.: Über die Theorie der einfachen Ungleichungen. J. Reine Angew. Math. **124**, 1–24 (1902)
9. Floyd, R.W.: Algorithm 97, shortest path. Commun. ACM **5**, 345 (1962)
10. Ford, L.R.: Network flow theory. Tech. Rep. P-923, The Rand Corp. (1956)
11. Gashkov, S.B., Sergeev, I.S.: A method for deriving lower bounds for the complexity of monotone arithmetic circuits computing real polynomials. Sb. Math. **203**(10), 1411–1147 (2012)
12. Grigoriev, D., Podolskii, V.V.: Tropical combinatorial Nullstellensatz and sparse polynomials. Found. Comput. Math. **20**, 753–781 (2020)
13. Held, M., Karp, R.M.: A dynamic programming approach to sequencing problems. SIAM J. Appl. Math. **10**, 196–210 (1962)
14. Jerrum, M., Snir, M.: Some exact complexity results for straight-line computations over semirings. J. ACM **29**(3), 874–897 (1982)
15. Jukna, S.: Lower bounds for tropical circuits and dynamic programs. Theory Comput. Syst. **57**(1), 160–194 (2015)
16. Jukna, S., Seiwert, H.: Approximation limitations of pure dynamic programming. SIAM J. Comput. **49**(1), 170–207 (2020)
17. Knuth, D.E.: The Art of Computer Programming, Vol. 3 - Sorting and Searching. Addison-Wesley, Boston (1973)
18. Levin, A.: Algorithm for the shortest connection of a group of graph vertices. Sov. Math. Dokl. **12**, 1477–1481 (1971)
19. Moore, E.F.: The shortest path through a maze. In: Proc. Internat. Sympos. Switching Theory, vol. II, pp. 285–292 (1957)
20. Mulmuley, K., Vazirani, U.V., Vazirani, V.V.: Matching is as easy as matrix inversion. Combinatorica **7**(1), 105–113 (1987)
21. Raz, R., Yehudayoff, A.: Multilinear formulas, maximal-partition discrepancy and mixed-sources extractors. J. Comput. Syst. Sci. **77**(1), 167–190 (2011)
22. Razborov, A.A.: Lower bounds on monotone complexity of the logical permanent. Math. Notes Acad. Sci. USSR **37**(6), 485–493 (1985)
23. Roy, B.: Transitivité at connexité. C. R. Acad. Sci. Paris **249**, 216–218 (1959). In French
24. Schnorr, C.P.: A lower bound on the number of additions in monotone computations. Theor. Comput. Sci. **2**(3), 305–315 (1976)
25. Schnorr, C.P.: Improved lower bounds on the number of multiplications/divisions which are necessary to evaluate polynomials. Theor. Comput. Sci. **7**, 251–261 (1978)
26. Schrijver, A.: Theory of Integer and Linear Programming. Wiley, New York (1986)
27. Schwartz, J.T.: Fast probabilistic algorithms for verification of polynomial identities. J. ACM **27**, 701–717 (1980)
28. Shamir, E., Snir, M.: On the depth complexity of formulas. Math. Syst. Theory **13**, 301–322 (1980)

29. Strassen, V.: Vermeidung von Divisionen. J. Reine Angew. Math. **264**, 184–202 (1973)
30. Strassen, V.: Polynomials with rational coefficients which are hard to compute. SIAM J. Comput. **3**(2), 128–149 (1974)
31. Ta-Shma, N.: A simple proof of the isolation lemma. Tech. rep., El Colloq. on Comput. Complexity (ECCC), Report Nr. 89 (2015)
32. Tiwari, P., Tompa, M.: A direct version of Shamir and Snir's lower bounds on monotone circuit depth. Inf. Process. Lett. **49**(5), 243–248 (1994)
33. Valiant, L.G.: Negation can be exponentially powerful. Theor. Comput. Sci. **12**, 303–314 (1980)
34. Viterbi, A.: Error bounds for convolutional codes and an asymptotically optimum decoding algorithm. IEEE Trans. Inf. Theory **13**(2), 260–269 (1967)
35. Warshall, S.: A theorem on boolean matrices. J. ACM **9**, 11–12 (1962)
36. Yehudayoff, A.: Separating monotone VP and VNP. In: Proc. of 51st Ann. ACM SIGACT Symp. on Theory of Computing, STOC, pp. 425–429. ACM, New York (2019)
37. Zippel, R.E.: Probabilistic algorithms for sparse polynomials. In: Proc. of EUROSAM 79. Lect. Notes in Comput. Sci., vol. 72, pp. 216–226. Springer, Berlin (1979)

# Chapter 2
# Combinatorial Bounds

**Abstract** As shown in the previous chapter, the tropical circuit complexity of homogeneous 0/1 optimization problems coincides with the Minkowski circuit complexity $\mathrm{L}(A)$ of sets $A \subseteq \{0, 1\}^n$ of their feasible solutions. In this chapter, we prove some general lower bounds of the form: if a set $A \subseteq \mathbb{N}^n$ of vectors has a specific *combinatorial* property, then $\mathrm{L}(A)$ is "large" and is near to $\mathrm{L}(A)$. We will show that so-called thin sets, including cover-free and Sidon sets, have such a property. Explicit constructions of these sets are also presented.

## 2.1 Cover-Free Sets

Recall (from Sect. 1.5) that a Minkowski $(\cup, +)$ circuit is a circuit which takes as inputs the single-element sets $\{\vec{0}\}, \{\vec{e}_1\}, \ldots, \{\vec{e}_n\}$ and produces some set $A \subseteq \mathbb{N}^n$ by using the set-theoretic union $(\cup)$ and Minkowski sum $(+)$ operations as gates. The size is the total number of gates in the circuit. As before, let

$$\mathrm{L}(A) := \text{min size of a Minkowski } (\cup, +) \text{ circuit producing the set } A.$$

Even if a set $A$ is very large, its Minkowski circuit complexity $\mathrm{L}(A)$ can be small. For example, let $A \subseteq \{0, 1\}^n$ be the set of all vectors with exactly $k$ 1-entries. Then $|A| = \binom{n}{k}$ but $\mathrm{L}(A) = \mathcal{O}(kn)$ (see Example 1.6 in Sect. 1.4). So, what sets $A$ require large Minkowski circuits, of size near to the total number $|A|$ of vectors in $A$? In this chapter, we present several combinatorial properties of sets $A \subseteq \mathbb{N}^n$ forcing $\mathrm{L}(A)$ to be large.

A set $A \subseteq \mathbb{N}^n$ is *cover-free* if the following holds for any vectors $a, b, c \in A$:

$$\text{if } a + b \geqslant c \text{ then } c \in \{a, b\}.$$

These sets were first introduced in 1964 by Kautz and Singleton [11] to investigate nonrandom superimposed binary codes. Cover-free sets have been considered for many cryptographic problems. Erdős et al. [4] have shown that the maximum

© The Author(s), under exclusive license to Springer Nature Switzerland AG 2023
S. Jukna, *Tropical Circuit Complexity*, SpringerBriefs in Mathematics,
https://doi.org/10.1007/978-3-031-42354-3_2

number $m = |A|$ of vectors in a cover-free set $A \subseteq \{0, 1\}^n$ satisfies $1.134^n < m < 1.25^n$.

One of the first general lower bounds for the monotone arithmetic circuit complexity of polynomials was proved by Schnorr [19] already in 1976:

(∗)  If $A \subseteq \mathbb{N}^n$ is cover-free, then $\texttt{Arith}(f) \geq |A| - 1$ holds for every polynomial
      $f$ whose set of exponent vectors is $A$.

(Cover-free sets are called *separated* sets in [19]). Together with Lemma 1.14, this gives the same lower bound on the Minkowski circuit complexity $\texttt{L}(A)$ of cover-free sets $A$.

**Theorem 2.1 (Schnorr [19])** *If $A \subseteq \mathbb{N}^n$ is cover-free, then $\texttt{L}(A) \geq |A| - 1$.*

In fact, Schnorr [19] proved a stronger result: the lower bound holds even on the number of union ($\cup$) gates.

The proof of (∗) in [19] uses the "method of inductive substitution," a kind of "gate elimination" argument used for Boolean circuits. Instead of presenting the original proof of Theorem 2.1 (as well as of the following Theorems 2.6 and 2.14), we will show that a slightly weaker lower bound $\texttt{L}(A) \geq |A|/2$ can be obtained by a fairly simple (and, apparently, more intuitive) "bottleneck counting" argument (see Theorem 2.18 in Sect. 2.4). Moreover, this latter bound holds also when besides only unit vectors $\vec{e}_i$, *any* vectors $x \in \mathbb{N}^n$ are allowed as inputs (are "given for free").

Let us give some applications of Theorem 2.1.

The *tropical matrix product problem* $\text{MP}_n$ is, given two $n \times n$ matrices $X = (x_{i,j})$ and $Y = (y_{i,j})$ with nonnegative entries, to compute $\min_{k \in [n]} \{x_{i,k} + y_{k,j}\}$ for all pairs of vertices $i, j \in [n]$. The problem $\text{MP}_n$ can be solved by a trivial $(\min, +)$ circuit of size $\mathcal{O}(n^3)$ (with $\binom{n}{2}$ output gates), and this circuit is almost optimal.

**Corollary 2.2** *Every $(\min, +)$ circuit solving the tropical matrix product problem $\text{MP}_n$ must have $\Omega(n^3)$ gates.*

This lower bound was earlier proved by Kerr [12, Theorem 6.1] using different arguments.

**Proof** Take a set of $n^2$ new variables, and consider the (arithmetic) polynomial

$$P_n = \sum_{i,j \in [n]} z_{i,j} \sum_{k \in [n]} x_{i,k} y_{k,j} = \sum_{i,j,k \in [n]} z_{i,j} x_{i,k} y_{k,j} \, .$$

Let $A$ be the set of all $|A| = n^3$ exponent vectors of $P_n$. Every monomial $x_{i,k} y_{k,j} z_{i,j}$ of $P_n$ is uniquely determined by any two of its variables. So, the set $A$ is cover-free, and Theorem 2.1 gives a lower bound $\texttt{L}(A) \geq |A| - 1 = n^3 - 1$. Since the set $A$ is homogeneous (of degree 3), Lemma 1.34 yields a lower bound $\texttt{Min}(A) = \texttt{L}(A) \geq n^3 - 1$ on the minimum size of a $(\min, +)$ circuit solving the problem $f_n = \min_{i,j \in [n]} \{z_{i,j} + \min_{k \in [n]} \{x_{i,k} + y_{k,j}\}\}$. Given a $(\min, +)$ circuit for $\text{MP}_n$, we can solve the problem $f_n$ by adding at most $\mathcal{O}(n^2)$ new gates. Thus, every $(\min, +)$ circuit solving the $\text{MP}_n$ problem must have at least $\texttt{Min}(A) - \mathcal{O}(n^2) = \Omega(n^3)$ gates.                                                                              □

Recall (from Example 1.8) that the *all-pairs shortest path* problem APSP($n$) is, given an assignment of nonnegative weights to the edges of the complete graph $K_n$ on $\{1, \ldots, n\}$, to compute the minimum weights of simple paths between all pairs of vertices. As shown in Example 1.8, this problem can be solved by a (min, +) circuit of size $\mathcal{O}(n^3)$ resulting from the Floyd–Warshall–Roy pure DP algorithm. The following consequence of Corollary 2.2 shows that this DP algorithm is essentially optimal (among all pure DP algorithms).

**Corollary 2.3** *Any* (min, +) *circuit solving the* APSP($n$) *problem must have* $\Omega(n^3)$ *gates.*

**Proof** Let $n = 3m$, and partite the set $[n]$ of vertices of $K_n$ into three disjoint sets $V_1$, $V_2$, and $V_3$, each of size $m$. Given a (min, +) circuit $\Phi$ solving APSP($n$), we can obtain a (min, +) circuit solving MP$_m$ by giving fixed weigh $+\infty$ to all edges of $K_n$, except those lying in $V_1 \times V_2$ or in $V_2 \times V_3$. Thus, by Corollary 2.2, the circuit $\Phi$ must have $\Omega(m^3) = \Omega(n^3)$ gates. $\qquad\square$

The *k-clique polynomial* is the following multilinear polynomial of $\binom{n}{2}$ variables:

$$\mathrm{CL}_{n,k}(x) = \sum_{\substack{S \subseteq [n] \\ |S| = k}} \prod_{\substack{i,j \in S \\ i < j}} x_{i,j} .$$

Note that, on every 0-1 vector $x$, $\mathrm{CL}_{n,k}(x)$ is the number of $k$-cliques in the subgraph $G_x$ of $K_n$ specified by vector $x$.

**Corollary 2.4** *For every subset* $A \subseteq \{0, 1\}^{\binom{n}{2}}$ *of exponent vectors of* $\mathrm{CL}_{n,k}$, *we have* $\mathrm{Min}(A) = \mathrm{Max}(A) = \mathrm{L}(A) \geqslant |A| - 1$.

**Proof** The set $A$ is homogeneous: each vector $a = (a_{i,j})$ in $A$ has exactly $\binom{k}{2}$ 1s corresponding to the edges of some $k$-clique. So, Lemma 1.34 yields the equalities $\mathrm{Min}(A) = \mathrm{Max}(A) = \mathrm{L}(A)$, and it remains to show that $\mathrm{L}(A) \geqslant |A| - 1$. By Theorem 2.1, it is enough to show that the set $A$ is cover-free. For this, assume the opposite, i.e., that the union of some two $k$-cliques contains some third $k$-clique. Since each $k$-clique has the same number $k$ of nodes, the latter clique must then have a node $u$ not in the first clique and a node $v$ not in the second clique. If $u = v$, then the node $u$ is not covered, and if $u \neq v$, then the edge $\{u, v\}$ is not covered by the union of the first two cliques, a contradiction. Thus, $A$ is cover-free. $\qquad\square$

The following corollary shows that, for non-homogeneous polynomials, tropical circuits can already be even exponentially smaller than Minkowski and monotone arithmetic circuits. Consider the following non-homogeneous polynomial:

$$\mathrm{SQ}_n(x) = \sum_{\emptyset \neq S \subseteq [n]} \prod_{i,j \in S} x_{i,j} . \tag{2.1}$$

That is, $\mathrm{SQ}_n$ is a (non-homogeneous) version of $\mathrm{CL}_{n,k}$ with the cardinality constraint "$|S| = k$" omitted and $i = j$ allowed.

**Corollary 2.5** *For the set $A \subseteq \{0, 1\}^{n \times n}$ of exponent vectors of* $SQ_n$, *both* $\text{Min}(A)$ *and* $\text{Max}(A)$ *are at most* $n^2$, *but* $L(A) \geqslant n^{\Omega(\sqrt{n})}$.

Using so-called shadows of Newton polytopes of polynomials, Hrubeš and Yehudayoff [7] showed a stronger lower bound $\text{Arith}(f) = 2^{\Omega(n)}$ for $f = SQ_n$.

***Proof*** The tropical versions $\min_{S \subseteq [n]} \sum_{i, j \in S} x_{i,j}$ and $\max_{S \subseteq [n]} \sum_{i, j \in S} x_{i,j}$ of the polynomial $SQ_n$ are trivially computable as $\Phi_1 = \min\{x_{i,j} : i, j \in [n]\}$ and as $\Phi_2 = \sum_{i, j \in [n]} x_{i,j}$ (input weights are nonnegative).

To show the lower bound $L(A) \geqslant n^{\Omega(\sqrt{n})}$, let $r = k^2$ for $k = \lfloor \sqrt{n/2} \rfloor$, and consider the $r$th envelope $A^{\langle r \rangle} = \{a \in A : a_1 + \cdots + a_n = r\}$ of the set $A$. By the homogenization lemma (Lemma 1.36), we have $L(A) \geqslant L(A^{\langle r \rangle})/(r+1)^2$. By Theorem 2.1, it remains to show that the set $A^{\langle r \rangle}$ is cover-free: then $L(A^{\langle r \rangle}) \geqslant |A^{\langle r \rangle}| - 1 = \binom{n}{k} - 1 \geqslant (n/k)^k - 1$, as desired.

The vectors in $A^{\langle r \rangle}$ correspond to the sets $S \times S$ of points in the grid $[n] \times [n]$ defined by $k$-element subsets $S$ of $[n]$. So, to show that $A$ is cover-free, it is enough to show that the inclusion $R \times R \subseteq (S \times S) \cup (T \times T)$ cannot hold for any $k$-element subsets $S, T$, and $R$ of $[n]$ such that $R \notin \{S, T\}$. Since all three sets have the same number $k$ of elements, $R \neq S$ gives an element $x \in R \setminus S$, and $R \neq T$ gives an element $y \in R \setminus T$. But then the point $(x, y)$ of $R \times R$ does not belong to $(S \times S) \cup (T \times T)$, as desired. Thus, $A^{\langle r \rangle}$ is cover-free.                                        □

## 2.2  Sidon Sets

For every set $A \subseteq \mathbb{N}^n$, the number $|A + A|$ of distinct vectors in the sumset $A + A$ is at most the number of all sets $\{a, b\}$ with $a, b \in A$, which is at most the number $\binom{|A|}{2}$ of sets with $a \neq b$ plus the number $|A|$ of sets with $a = b$. So, for every set $A \subseteq \mathbb{N}^n$, we have

$$|A + A| \leqslant \binom{|A|}{2} + |A| = \frac{|A|(|A| - 1)}{2} + |A| = \frac{|A|(|A| + 1)}{2} = \binom{|A| + 1}{2}.$$

Sidon sets[1] $A \subseteq \mathbb{N}^n$ are those for which the equality $|A + A| = \binom{|A|+1}{2}$ holds: all sums of pairs of (not necessarily distinct) vectors in $A$ are distinct. That is, if we know the sum of two vectors of $A$, we know which vectors were added. Formally, a set $A \subseteq \mathbb{N}^n$ is a *Sidon set* if the following holds for all vectors $a, b, c, d \in A$:

$$\text{if } a + b = c + d \text{ then } \{c, d\} = \{a, b\}.$$

For example, in the dimension $n = 1$, $A = \{1, 2, 5, 7\}$ is a Sidon set, but $B = \{1, 2, 4, 5, 7\}$ is not a Sidon set because, for example, $2 + 4 = 1 + 5$. Sidon sets $A \subseteq \mathbb{N}$ are also known as *Golomb rulers*.

---

[1] The term "Sidon sets" was coined by Paul Erdős in honor of Fourier analyst Simon Sidon [20] who introduced these sets in order to solve a problem in harmonic analysis.

Recall that a set $A \subseteq \mathbb{N}^n$ is cover-free iff for every $a, b, c \in A$ and $d \in \mathbb{N}^n$ (not necessarily $d \in A$), $a + b = c + d$ implies $c \in \{a, b\}$. Thus,

- Every cover-free set is also a Sidon set.

But there are Sidon sets which are not cover-free. For example, Sidon set $A$ constructed in Example 2.8 below is clearly not cover-free because it is even not an antichain: say, the all-0 vector belongs to it.

The following special case of a more general lower bound proved by Gashkov [5] extends Schnorr's bound to Sidon sets.

**Theorem 2.6 (Gashkov [5])** *If $A \subseteq \mathbb{N}^n$ is a Sidon set, then $L(A) \geqslant |A| - 1$.*

As in Schnorr's bound for cover-free sets (Theorem 2.1), Gashkov's bound is already on the number of union ($\cup$) gates.

It is know that $|A| \leqslant 2^{3n/5}$ holds for every Sidon set $A \subseteq \{0, 1\}^n$ [2, 17]. Constructions of large Sidon sets are also known.

**Example 2.7 (Greedy Construction)** A Sidon set $A \subseteq \{0, 1\}^n$ of size $|A| \geqslant 2^{n/3}$ can be constructed in a "greedy manner." Start with an empty set $A = \emptyset$, and at each step, include a vector $x$ in $A$ if $x$ is not of the form $x = a + b - c$ for some vectors $a, b$, and $c$ already in the set $A$. If we have already constructed a set $A$ of size $|A| = m$ and if $m + 3\binom{m}{3} < 2^n$, then at least one new vector $x$ will be included in $A$ (each triple of the vectors in $A$ "kills" at most 3 of the remaining vectors). So, since this inequality holds for $m := 2^{n/3}$, the final Sidon set will have $|A| \geqslant 2^{n/3}$ vectors.  □

An *explicit* Sidon set $A \subseteq \{0, 1\}^n$ of size $|A| = 2^{n/2}$ is known.

**Example 2.8 (Cubic Parabola, Lindström [16])** Let $n = 2m$, and identify the elements of the field $GF(2^m)$ with vectors in $\{0, 1\}^m$. The (graph of the) cubic parabola over $\mathbb{F}$ is the set $A = \{(a, a^3) : a \in \{0, 1\}^m\} \subseteq \{0, 1\}^n$. As customary, we view 0-1 vectors $a = (a_1, \ldots, a_m)$ as coefficient vectors of univariate polynomials $p_a(z) = a_1 + a_2 z + a_3 z^2 + \cdots + a_m z^{m-1}$ of degree at most $m - 1$ over $GF(2)$ when rising them to a power; as the multiplication can end up with a polynomial of degree greater than $m - 1$, we take the result modulo an irreducible polynomial.

To show that $A$ is a Sidon set, fix any two vectors $c, d \in \{0, 1\}^m$, and consider the equation $(x, x^3) + (y, y^3) = (c, c^3) + (d, d^3)$. It is enough to show that this equation has at most one unordered pair $\{x, y\}$ of 0-1 solutions over the semigroup $(\mathbb{N}^n, +)$. If $c = d$, then there is only one solution $\{x, y\}$ of $x + y = c + d = 2c$ with $x = y = c$: vectors $x, y$, and $c$ are 0-1 vectors, and the addition is over $(\mathbb{N}^n, +)$. So, assume that $c \neq d$. It is enough to show that then the equation cannot have more than one solution $\{x, y\}$ even over the field $GF(2^n)$.

The equation $(x, x^3) + (y, y^3) = (c, c^3) + (d, d^3)$ is equivalent to the system of two equations $x + y = a$ and $x^3 + y^3 = b$ with $a := c + d \neq 0$ and $b := c^3 + d^3$. Since we are now working over a field of characteristic 2, the identity $(x + y)^3 = x^3 + y^3 + 3xy(x + y)$ turns into $(x + y)^3 = x^3 + y^3 + xy(x + y) = x^3 + y^3 + axy$. Thus, $axy = (x + y)^3 + (x^3 + y^3) = a^3 + b$, meaning that $x$ and $y$ must satisfy

$x + y = a \neq 0$ and $xy = a^2 + b/a$. By Vieta's formula[2], $x$ and $y$ are then the roots of the quadratic polynomial $aX^2 - a^2X + (a^3 + b)$, and there can be only one pair of them.                                                                                      $\square$

**Corollary 2.9** *For the cubic parabola $A \subseteq \{0, 1\}^n$, we have $L(A) \geqslant 2^{n/2} - 1$.*

Sidon sets $A \subseteq \{0, 1\}^m$ are not necessarily homogeneous, so that the same lower bounds on the Minkowski circuit complexity $L(A)$ do not necessarily extend, via Lemma 1.34, to the *tropical* circuit complexity of optimization problems on these sets. But every set of 0-1 vectors can be made homogeneous by doubling the length of vectors. Namely, let $n = 2m$ and define the *homogeneous extension* of a set $A \subseteq \{0, 1\}^m$ to be the set $B = \{(a, \underline{a}) : a \in A\} \subseteq \{0, 1\}^n$, where $\underline{a}$ denotes the componentwise negation of a 0-1 vector $a$. For example, if $a = (0, 0, 1, 0, 1)$, then $\underline{a} = (1, 1, 0, 1, 0)$. Note that the set $B$ is already homogeneous because each of its vectors has exactly $m$ ones. It is not difficult to see that homogeneous extensions of Sidon sets are also Sidon sets. So, together with Lemma 1.34, Corollary 2.9 yields the following lower bound on the tropical circuit complexity of optimization problems on homogeneous extensions of cubic parabolas.

**Corollary 2.10** *Let $n = 4m$, and $A = \{(a, a^3, \underline{a}, \underline{a}^3) : a \in \{0, 1\}^m\} \subseteq \{0, 1\}^n$. Then $\mathrm{Min}(A) = \mathrm{Max}(A) \geqslant 2^{n/4} - 1$.*

### Sidon Sets and Null-Chain Free Boolean Circuits

Lower bounds of the form $L(A) \geqslant 2^{cn}$ with constants $0 < c < 1$ for explicit sets $A \subseteq \{0, 1\}^n$ were proved using various arguments (and arguments often matter even more than resulting numerical bounds themselves):

o  Discrepancy arguments (Raz and Yehudayoff [18])
o  Expander graphs (Jukna [9])
o  Communication complexity arguments (Srinivasan [21])
o  Error-correcting codes (Cavalar et al. [1])

Actually, a lower bound of the form $L(A) \geqslant 2^{n/6}$ was also implicit in a paper [14] by Kuznetsov from 1981 concerning (non-monotone) Boolean circuits without null-chains. Let us show how does this happen.

A *DeMorgan circuit* is a Boolean circuit over $\{\vee, \wedge, \neg\}$ with negations applied only to the inputs. That is, this is a circuit over $\{\vee, \wedge\}$ whose inputs are the variables $x_1, \ldots, x_n$ and their negations $\bar{x}_1, \ldots, \bar{x}_n$. It is well known (see, e.g., [22, p. 195]) that every Boolean $\{\vee, \wedge, \neg\}$ has an equivalent DeMorgan circuit of at most double size, so the restriction to tight negations is actually not a real restriction. Namely, we double all AND and OR gates: one output of a pair is negated and the other one not; after that we can apply the DeMorgan rules without increasing the number of gates.

---

[2] If $\alpha$ and $\beta$ are the solutions of $c_2x^2 + c_1x + c_0 = 0$, then $\alpha + \beta = -c_1/c_2$ and $\alpha\beta = c_0/c_2$.

Every DeMorgan circuit $\phi$ not only computes a unique Boolean function but also produces (purely syntactically) a unique formal DNF $D_\phi$. When forming this DNF, we allow the laws $x \wedge x = x$ (idempotence) and $x \vee xy = x$ (absorption) to be used, but the law $x \wedge \bar{x} = 0$ (cancelation) is not used. So, in general, the produced DNF may contain zero terms, i.e., those containing a variable together with its negation. A circuit $\phi$ is *null-chain free* if $D_\phi$ contains no zero terms.

Associate with every set $A \subseteq \{0, 1\}^n$ the Boolean function $f_A$ such that $f_A(x) = 1$ iff $x \in A$. Kuznetsov [14, Lemma 3] proved (without using the term "Sidon set") the following lower bound.

**Lemma 2.11 (Kuznetsov [14])** *Let $A \subseteq \{0, 1\}^n$ be a Sidon set, any two vectors of which differ in at least two positions. Then every null-chain free DeMorgan circuit computing $f_A$ has at least $|A|$ gates.*

Igor Sergeev (private communication) observed that this lemma yields lower bounds of the form $L(A) \geqslant 2^{n/6}$ for explicit sets $A \subseteq \{0, 1\}^n$. Namely, associate with a set $A \subseteq \{0, 1\}^n$ the following multilinear polynomial of $2n$ variables:

$$P_A(x, y) = \sum_{a \in A} \prod_{i:a_i=1} x_i \prod_{i:a_i=0} y_i \,.$$

**Proposition 2.12** *If the polynomial $P_A(x, y)$ can be computed by a monotone arithmetic circuit of size $s$, then the Boolean function $f_A(x)$ can be computed by a null-chain free DeMorgan circuit of the same size $s$.*

*Proof* Take a monotone arithmetic $(+, \times)$ circuit $\Phi(x, y)$ computing the polynomial $P_A(x, y)$. By Lemma 1.31, the circuit $\Phi$ also (syntactically) produces that polynomial. Turn the circuit $\Phi$ into a Boolean circuit $\phi(x, y)$: replace $+$ gates by $\vee$ gates and $\times$ gates by $\wedge$ gates. The formal DNF produced by the resulting Boolean circuit $\phi$ is the Boolean version

$$D_\phi(x, y) = \bigvee_{a \in A} \bigwedge_{i:a_i=1} x_i \bigwedge_{i:a_i=0} y_i$$

of the polynomial $P_A(x, y)$. Note that no monomial of $D_\phi(x, y)$ contains both variables $x_i$ and $y_i$. So, if we replace every input variable $y_i$ in $\phi$ by $\bar{x}_i$, then the DNF $D_{\phi'}(x)$ produced by the DeMorgan circuit $\phi'(x) = \phi(x, \bar{x})$ has no zero terms, meaning that the obtained circuit $\phi'$ is null-chain free. This circuit accepts an input vector $x \in \{0, 1\}^n$ iff $x \in A$. $\square$

To construct a Sidon set $A \subseteq \{0, 1\}^n$ of size $|A| \geqslant 2^{n/3}$ with any two vectors of $A$ at Hamming distance at least two, Kuznetsov [14] used a greedy construction similar to that used in Example 2.7. Instead of all vectors, consider only vectors with an even number of ones (to ensure that every two vectors differ in at least two positions), and replace the "increase condition" $m + 3\binom{m}{3} < 2^n$ by $m + 3\binom{m}{3} < 2^{n-1}$. So, his result also yields a lower bound $L(A) \geqslant 2^{n/3}$ on the size of monotone arithmetic circuits computing a multilinear polynomial $P_A$ of $2n$ variables.

## 2.3 Thin Sets

A natural generalization of Sidon sets is that of $(k, l)$-thin sets introduced by Erdős and Harzheim in [3]. Let $1 \leqslant k \leqslant l$ be integers. A set $A \subseteq \mathbb{N}^n$ of vectors is $(k, l)$-*thin* if the following holds for any two subsets $X, Y \subseteq \mathbb{N}^n$ of vectors:

$$\text{if } X + Y \subseteq A \text{ then } |X| \leqslant k \text{ or } |Y| \leqslant l.$$

In other words, a set $A \subseteq \mathbb{N}^n$ is $(k, l)$-thin if for any $k + 1$ distinct vectors $\vec{a}_1, \ldots, \vec{a}_{k+1} \in \mathbb{N}^n$, the system of relations $\vec{a}_1 + x \in A, \ldots, \vec{a}_{k+1} + x \in A$ has at most $l$ distinct solutions $x \in \mathbb{N}^n$. The interpretation of thin sets $A \subseteq \mathbb{N}^n$ in terms of *graphs* is the following. Associate with $A$ the (infinite) bipartite graph $G_A = \{(x, y) \in \mathbb{N}^{n \times n} : x + y \in A\}$. Then $A$ is $(k, l)$-thin iff $G_A$ contains no copy of a complete bipartite $(k + 1) \times (l + 1)$ graph as a subgraph. It is not difficult to show that $(1, 1)$-thin sets are exactly Sidon sets.

**Proposition 2.13** *A set $A \subseteq \mathbb{N}^n$ is $(1, 1)$-thin iff $A$ is a Sidon set.*

**Proof** ($\Leftarrow$) Suppose that $A$ is not $(1, 1)$-thin. Then there are vectors $a = x + y$ and $b = x' + y'$ in $A$ for some vectors $x \neq x'$ and $y \neq y'$ in $\mathbb{N}^n$ such that the vectors $c = x + y'$ and $d = x' + y$ also belong to $A$. Then clearly $a + b = c + d$ but neither $c = a$ nor $c = b$ can hold, meaning that $A$ is not a Sidon set.

($\Rightarrow$) Suppose that a set $A \subseteq \mathbb{N}^n$ is not a Sidon set. Hence, there are vectors $a, b, c, d \in A$ such that $a + b = c + d$ but $c \notin \{a, b\}$. To show that then $A$ is not $(1, 1)$-free, it is enough to show that there exist vectors $x \neq x'$ and $y \neq y'$ in $\mathbb{N}^n$ such that $\{x, x'\} + \{y, y'\} \subseteq A$. We define the desired vectors componentwise. If $a_i < c_i$, then take $x_i = 0$, $x_i' = c_i - a_i$, $y_i = a_i$, and $y_i' = d_i$ ($= a_i - c_i + b_i$). If $a_i \geqslant c_i$, then take $x_i = a_i - c_i$, $x_i' = 0$, $y_i = c_i$, and $y_i' = b_i$ ($= c_i - a_i + d_i$). In both cases, all four vectors $x + y = a$, $x' + y' = b$, $x + y' = d$, and $x' + y = c$ belong to $A$. Moreover, $c \neq a$ implies $x \neq x'$, and $c \neq b$ implies $y \neq y'$. □

**Theorem 2.14 (Gashkov and Sergeev [5, 6])** *Let $1 \leqslant k \leqslant l$ be integers and $A \subseteq \mathbb{N}^n$ be a finite set of vectors. If $A$ is $(k, l)$-thin, then $L(A) \geqslant |A| / \max\{k^3, l^2\} - 1$.*

In the case when $k = l$, this lower bound was proved by Gashkov [5] and was later extended by Gashkov and Sergeev [6] to any $k \leqslant l$. An important aspect of the "asymmetric extension" (giving meaningful lower bounds also when $k \neq l$) is that, together with known explicit constructions of large $(k, l)$-thin sets $A \subseteq \{0, 1\}^n$, Theorem 2.14 yields the *highest* known explicit lower bound $L(A) \geqslant 2^{n-o(n)}$ on the Minkowski circuit complexity and hence (according to Corollary 2.16) also on the monotone arithmetic circuit complexity of any multilinear polynomial $f(x) = \sum_{a \in A} c_a \prod_{i=1}^n x_i^{a_i}$ with coefficients $c_a > 0$.

**Example 2.15 (Norm Sets)** Let $q$ be a prime power $t \geqslant 2$ an integer, and consider the field $\mathbb{F} = GF(q^t)$ with $q^t$ elements. The *norm function* is a mapping $N : GF(q^t) \rightarrow GF(q)$ given by $N(a) = a \cdot a^q \cdots a^{q^{t-1}} = a^{(q^t-1)/(q-1)}$. Consider the set $A = \{a \in \mathbb{F} : N(a) = 1\}$ of all elements of unit norm. It is known (see, e.g., [15]) that $|A| = (q^t - 1)/(q - 1)$. Kollár, Rónyai, and Szabó [13, Theorem 3.3] proved that, for every $t$ distinct elements $a_1, \ldots, a_t$ of $\mathbb{F}$, the system of equations

$$N(a_1 + x) = 1, N(a_2 + x) = 1, \ldots, N(a_t + x) = 1$$

has at most $t!$ solutions $x \in \mathbb{F}$. Hence, the set $A$ is $(t, t!)$-thin over the group $(\mathbb{F}, +)$. Now let $q = 2^r$ and $n = rt$. By viewing elements of $GF(2^n)$ as vectors in $\{0, 1\}^n$, we obtain an explicit set $A_{n,t} \subseteq \{0, 1\}^n$ of $|A_{n,t}| = (2^{rt} - 1)/(2^r - 1) \geqslant 2^{r(t-1)} = 2^{n-n/t}$ vectors, which is $(t, t!)$-thin over $(\mathbb{F}, +)$ and, hence, also over the semigroup $(\mathbb{N}^n, +)$. $\square$

**Corollary 2.16** *For every norm set $A_{n,t} \subseteq \{0, 1\}^n$, we have $\mathrm{L}(A_{n,t}) \geqslant 2^{n-n/t-2t \log t}$. In particular, $\mathrm{L}(A_{n,\sqrt{n}}) \geqslant 2^{n-2\sqrt{n} \log n}$.*

**Proof** By the construction in Example 2.15, we know that the set $A_{n,t}$ is $(t, t!)$-thin and has $|A_{n,t}| = 2^{n-n/t}$ vectors. Theorem 2.14 yields the lower bound $\mathrm{L}(A_{n,t}) \geqslant 2^{n-n/t}/(t!)^2 \geqslant 2^{n-n/t-2t \log t}$. For $t = \sqrt{n}$, we obtain $\mathrm{L}(A_{n,t}) \geqslant 2^{n-2\sqrt{n} \log n}$. $\square$

Recall that the *homogeneous extension* of a set $A \subseteq \{0, 1\}^m$ is the set $B = \{(a, \underline{a}): a \in A\} \subseteq \{0, 1\}^n$, where $\underline{a}$ denotes the componentwise negation of a 0-1 vector $a$. It is easy to verify that if $A$ is $(k, l)$-thin, then also $B$ is $(k, l)$-thin. So, together with the construction of norm-sets (Example 2.15), Theorem 2.14 gives the highest currently known explicit lower bound on the tropical circuit complexity of optimization problems.

**Corollary 2.17** *Let $n = 2m$, where $m$ is a square of an integer, and let $A \subseteq \{0, 1\}^n$ be the homogeneous extension of the norm-set $A_{m,t} \subseteq \{0, 1\}^m$ for $t = \sqrt{m}$. Then $\mathrm{Min}(A) \geqslant 2^{n/2-o(n)}$ and $\mathrm{Max}(A) \geqslant 2^{n/2-o(n)}$.*

## 2.4 The Bottleneck Counting Argument

We now turn to the proof of the promised lower bound on the size of Minkowski circuits. Numerically, the bound is slightly worse than those given by Theorems 2.1, 2.6, and 2.14, but it holds for more general circuits, and (perhaps, more importantly) the proof is fairly elementary.

Recall that a Minkowski circuit can only use $n + 1$ singletons $\{\vec{0}\}, \{\vec{e}_1\}, \ldots, \{\vec{e}_n\}$ as inputs. In a $k$-*extended* Minkowski $(\cup, +)$ circuit, we allow *any* sets $X \subseteq \mathbb{N}^n$ of $|X| \leqslant k$ to be used as inputs. In the context of *arithmetic* circuits, this means that any polynomials with at most $k$ monomials are allowed to be used as inputs (for free).

For a set $A \subseteq \mathbb{N}^n$, let $\mathrm{L}_k(A)$ denote the minimum number of gates in a $k$-extended Minkowski $(\cup, +)$ circuit producing the set $A$. Note that $\mathrm{L}(A) \geqslant \mathrm{L}_k(A)$ holds for all $k \geqslant 1$, and the gap can be large already for $k = 1$. As a trivial example, take a vector $a \in \mathbb{N}^n$ with $a_1 = 2^{2^n}$ and $a_i = 0$ for $i = 2, \ldots, n$, then $\mathrm{L}_1(\{a\}) = 0$ but $\mathrm{L}(\{a\}) \geqslant \log_2 a_1 = 2^n$, because Minkowski sum gates have fan-in only 2. Also, we have $\mathrm{L}_1(A) \leqslant |A| - 1$ for every finite set $A \subseteq \mathbb{N}^n$: just take all singletons $\{a\}$ for $a \in A$ as inputs, and compute their union.

**Theorem 2.18 (Jukna [10])** *Let* $1 \leqslant k \leqslant l$, *and let* $A \subseteq \mathbb{N}^n$ *be a finite set of* $|A| > k$ *vectors. If $A$ is* $(k, l)$-*thin, then* $L_k(A) \geqslant |A|/2lk^2$.

In particular, since Sidon sets are $(1, 1)$-thin (Proposition 2.13), we have quite tight bounds $|A|/2 \leqslant L_1(A) \leqslant |A| - 1$ for every Sidon set $A \subseteq \mathbb{N}^n$. For example, if $A \subseteq \{0, 1\}^n$ is the cubic parabola (see Example 2.8), then $2^{n/2-1} \leqslant L_1(A) \leqslant 2^{n/2} - 1$.

The proof idea of Theorem 2.18 is to analyze the "progress" made along the input–output paths of a circuit until particular "bottlenecks" are found. Namely, given a finite set $A \subseteq \mathbb{N}^n$ of vectors, we take an arbitrary $k$-extended Minkowski $(\cup, +)$ circuit $\Phi$ producing the set $A$ and argue as follows, where $E$ is the set of edges in the underlying graph of $\Phi$:

(i) Associate with each edge $e \in E$ a special subset $A_e \subseteq A$, the "content" of $e$.
(ii) Show that for every vector $a \in A$ there is an edge $e \in E$ such that $a \in A_e$ and $|A_e| \leqslant k^2 l$.

If the circuit $\Phi$ has $t$ gates, then it has $|E| \leqslant 2t$ edges (the fan-in of gates is 2). On the other hand, by (ii), there must be $|E| \geqslant |A|/k^2 l$ edges and, hence, $t \geqslant |A|/2k^2 l$ gates in the circuit $\Phi$.

To implement Step (i), let a (combinatorial) *rectangle*[3] be a Minkowski sum $X + Y$ of two nonempty sets of vectors $X, Y \subseteq \mathbb{N}^n$; we prefer to use the term "rectangle" instead of "sumset" only in order to be able to refer to the parts $X$ and $Y$ of this sum *set*. Rectangles naturally emerge in Minkowski circuits (as well as in circuits over any semiring) as "contents" of their gates and edges.

Let $\Phi$ be a $k$-extended Minkowski $(\cup, +)$ circuit and $A \subseteq \mathbb{N}^n$ be the set of vectors produced by it (at the output gate). Recall that the set $X_v$ *produced* at a gate $v$ entered from gates $u$ and $w$ is the union $X_v = X_u \cup X_w$ if $v = u \cup w$ and is the Minkowski sum $X_v = X_u + X_w$ if $v = u + w$. The set $X_v$ produced at an input node $v$ is the set $X_v \subseteq \mathbb{N}^n$ of $|X_v| \leqslant k$ vectors held by this node.

The set $X_v$ of vectors produced at a gate $v$ does not need to lie in $A$, but at least one of its translates $X_v + y = \{x + y : x \in X\}$ by some vector $y \in \mathbb{N}^n$ must already lie in $A$. The *residue*

$$Y_v := \{y \in \mathbb{N}^n : X_v + y \subseteq A\}$$

of $A$ at gate $v$ collects all such vectors $y$. In particular, if $X_v \subseteq A$, then $Y_v = \{\vec{0}\}$. Note that $X_v$ is the "real" part of the rectangle $X_v + Y_v$ (produced by the circuit at gate $v$), while $Y_v$ is only "imaginary" part (determined by the set $A$ produced at the end). Following Jerrum and Snir [8], we call the rectangle $X_v + Y_v$ the *content* of gate $v$. Note that contents of gates are subsets of the set $A$ produced by the entire circuit, that is, the inclusion $X_v + Y_v \subseteq A$ holds for every gate $v$. Intuitively, the

---

[3] This term has little till nothing to do with usual geometric rectangles. It is borrowed from communication complexity to reflect the fact that the set $X + Y$ of vectors is of a very special form: *every* vector of $X$ is added to *every* vector of $Y$.

content of a gate contains all vectors of $A$ to the production of which the gate $v$ "contributes." We also define the *content* of an edge $e = (u, v)$ to be the rectangle $X_v + Y_v$ if $v$ is a Minkowski sum gate and to be the rectangle $X_u + Y_v$ if $v$ is a union gate:

Note that the inclusion $X_u + Y_v \subseteq A$ holds also in this latter case (when $v$ is a union gate) because then $Y_v = Y_u \cap Y_w \subseteq Y_u$ and $X_u + Y_u \subseteq A$.

The reason for the asymmetry when defining the contents of edges entering different types of gates is to ensure the following "content propagation" property. Take an arbitrary vector $a \in A$ in the set $A$ produced by a Minkowski circuit. This vector belongs to the content $X_o + Y_o = A + \{\vec{0}\}$ of the output gate $o$. Start at the output gate $o$, and construct a path in the underlying directed acyclic graph by going backward and by always choosing an edge whose content contains our vector $a$. There may be many such paths, and the following lemma ensures that *none* of them will "get stuck," that is, every such path will eventually reach some input node.

Recall that for a directed edge $e = (u, v)$, which goes from $u$ to $v$, the node $u$ is the *tail* and $v$ is the *head* of $e$.

**Lemma 2.19 (Content Propagation Property)** *Let $v$ be a gate and $a \in X_v + Y_v$ be a vector in its content.*

(1) *If $v$ is a union ($\cup$) gate, then vector $a$ belongs to the content of at least one edge entering $v$, as well as to the content of the tail of that edge.*
(2) *If $v$ is a Minkowski sum ($+$) gate, then vector $a$ belongs to the contents of both edges entering $v$ and to the contents of the tails of these edges.*

**Proof** The gate $v$ is entered by edges from some two gates $u$ and $w$. Let first $v = u \cup w$ be a union gate. Since vector $a$ belongs to the content $X_v + Y_v = (X_u \cup X_w) + Y_v$ of the gate $v$, the vector $a$ must belong to at least one of the contents $X_u + Y_v$ or $X_w + Y_v$ of the edges entering gate $v$. Assume w.l.o.g. that $a$ belongs to $X_u + Y_v$. Since in the case of a union gate, we have that $Y_v = Y_u \cap Y_w$, vector $a$ belongs to the content $X_u + Y_u$ of the tail $u$ of edge $(u, v)$, as well.

Now let $v = u + w$ be a Minkowski sum gate. In this case, the contents of both edges $(u, v)$ and $(w, v)$ coincide with the content $X_v + Y_v$ of gate $v$ (by the definition of the contents of edges). So, vector $a$ (automatically) belongs to the contents of both these edges, and it remains to show that vector $a$ belongs to the contents $X_u + Y_u$ and $X_w + Y_w$ of the gates $u$ and $w$. Since $a \in X_v + Y_v = (X_u + X_w) + Y_v$, we have $a = x_u + x_w + y$ for some vectors $x_u \in X_u$, $x_w \in X_w$, and $y \in Y_v$. Since $y \in Y_v$, we know that the inclusion $X_u + X_w + y \subseteq A$ and, hence, also the inclusion $X_u + (x_w + y) \subseteq A$ hold. Hence, $x_w + y \in Y_u$ and, similarly, $x_u + y \in Y_w$. Thus, the vector $a = x_u + (x_w + y) = x_w + (x_u + y)$ belongs to $X_u + Y_u$ and to $X_w + Y_w$, as desired.                                                                            $\square$

**Proof of Theorem 2.18** Let $A \subseteq \mathbb{N}^n$ be a $(k, l)$-thin set of $|A| > k$ vectors. Take a Minkowski $(\cup, +)$ circuit $\Phi$ which can use arbitrary sets $X \subseteq \mathbb{N}^n$ of $|X| \leqslant k$ vectors as inputs and produces the set $A$. Our goal is to show that the circuit must have at least $|A|/2lk^2$ gates. Call an edge of the circuit $\Phi$ *light* if its content has at most $lk^2$ vectors.

**Claim 2.20** *Every vector $a \in A$ belongs to the content of at least one light edge.*

Note that this claim already yields Theorem 2.18. Indeed, the claim implies that the set $A$ is contained in the union of the contents of light edges. Thus, since each of these contents has $\leqslant k^2l$ vectors, there must be $r \geqslant |A|/k^2l$ (light) edges in the circuit and, hence, at least $r/2 = |A|/2lk^2$ gates because each gate has indegree only two.

To prove Claim 2.20, call a gate $u$ *cheap* if $|X_v| \leqslant k$ and *expensive* otherwise. Fix a vector $a \in A$. Start at the output gate of the circuit, and construct a path in the underlying directed acyclic graph by going backward and using the following rule, where $v$ is the last already reached gate.

1. If $v$ is a union $(\cup)$ gate, then follow an edge whose content contains vector $a$; if vector $a$ is contained in the contents of both edges entering $v$, then follow any one of these two edges.
2. If $v$ is a Minkowski sum $(+)$ gate, then go to either of the two inputs if they are both expensive or are both cheap, and go to the expensive input if the second input is cheap.

By the content propagation property (Lemma 2.19), the content of *every* edge along the entire path contains our vector $a$, and we will eventually reach some input gate. Since, by our assumption $|A| > k$, the output gate is expensive, and since each input gate is cheap, there must be an edge $e = (u, v)$ in this input–output path with the cheap tail $u$ and expensive head $v$, that is, with $|X_u| \leqslant k$ and $|X_v| > k$. If $v = u + w$ is a Minkowski sum gate, then, since the gate $u$ is cheap, step (2) in the construction of the path implies that the second gate $w$ entering $v$ must also be cheap, that is, $|X_w| \leqslant k$ must hold as well. Since $X_v + Y_v \subseteq A$ and $|X_v| > k$, the $(k, l)$-thinness of the set $A$ implies that $|Y_v| \leqslant l$ must hold. Thus, we have the following information about the found edge $e = (u, v)$:

$$|X_u| \leqslant k \qquad |X_w| \leqslant k \text{ if } v = u + w$$

$$u \circ \qquad \circ w$$

$$v \circ$$

$$|X_v| > k \text{ but } |Y_v| \leqslant l$$

It remains to show that the edge $e = (u, v)$ is a light edge. If $v = u \cup w$ is a union gate, then the content of edge $e$ is the rectangle $X_u + Y_v$. Thus, $|X_u + Y_v| \leqslant |X_u| \cdot |Y_v| \leqslant kl$, meaning that the edge $e$ is light. If $v = u + w$ is a Minkowski sum gate, then the content of edge $e$ is the rectangle $X_v + Y_v = X_u + X_w + Y_v$. Since in this case, we additionally have the upper bound $|X_w| \leqslant k$ on the number of vectors produced at the second gate $w$, we have $|X_v + Y_v| \leqslant |X_u| \cdot |X_w| \cdot |Y_v| \leqslant k^2 l$, meaning that the edge $e$ is light also in this case. □

**Remark 2.21** The proof (for $k = l = 1$) shows an interesting structural property of Minkowski circuits $\Phi$ producing Sidon sets $A \subseteq \mathbb{N}^n$: it follows from Claim 2.20 that then for every vector $a \in A$ there must be an edge in $\Phi$ whose content consists of only this vector $a$. That is, each vector of $A$ requires its "own" edge in the circuit. □

# References

1. Cavalar, B.P., Kumar, M., Rossman, B.: Monotone circuit lower bounds from robust sunflowers. In: LATIN 2020: Theoretical Informatics - 14th Latin American Symposium, Proceedings. Lect. Notes in Comput. Sci., vol. 12118, pp. 311–322. Springer, Cham (2020)
2. Cohen, G., Litsyn, S., Zémor, G.: Binary $b_2$-sequences: A new upper bound. J. Combin. Theory Ser. A **94**, 152–155 (2001)
3. Erdős, P., Harzheim, E.: Congruent subsets of infinite sets of natural numbers. J. Reine Angew. Math. **367**, 207–214 (1986)
4. Erdős, P., Frankl, P., Füredi, Z.: Families of finite sets in which no set is covered by the union of two others. J. Combin. Theory Ser. A **33**, 158–166 (1982)
5. Gashkov, S.B.: On one method of obtaining lower bounds on the monotone complexity of polynomials. Vestnik MGU, Ser. 1 Math. Mech. **5**, 7–13 (1987)
6. Gashkov, S.B., Sergeev, I.S.: A method for deriving lower bounds for the complexity of monotone arithmetic circuits computing real polynomials. Sb. Math. **203**(10), 1411–1147 (2012)
7. Hrubeš, P., Yehudayoff, A.: Shadows of Newton polytopes. In: 36th Comput. Compl. Conf. (CCC 2021). Leibniz Int. Proc. in Informatics, vol. 200, pp. 9:1–9:23 (2021)
8. Jerrum, M., Snir, M.: Some exact complexity results for straight-line computations over semirings. J. ACM **29**(3), 874–897 (1982)
9. Jukna, S.: Lower bounds for tropical circuits and dynamic programs. Theory Comput. Syst. **57**(1), 160–194 (2015)
10. Jukna, S.: Tropical complexity, Sidon sets and dynamic programming. SIAM J. Discrete Math. **30**(4), 2064–2085 (2016)
11. Kautz, W., Singleton, R.: Nonrandom binary superimposed codes. IEEE Trans. Inf. Theory **10**(4), 363–377 (1964)
12. Kerr, L.R.: The effect of algebraic structure on the computation complexity of matrix multiplications. Ph.D. thesis, Cornell Univ., Ithaca, NY (1970)
13. Kollár, J., Rónyai, L., Szabó, T.: Norm-graphs and bipartite Turán numbers. Combinatorica **16**(3), 399–406 (1996)
14. Kuznetzov, S.E.: Circuits composed of functional elements without zero paths in the basis $\{\&, \vee, -\}$. Izv. Vyssh. Uchebn. Zaved. Mat. **228**(5), 56–63 (1981). In Russian
15. Lidl, R., Niederreiter, H.: Introduction to Finite Fields and Their Applications. Cambridge University Press, Cambridge (1986)
16. Lindström, B.: Determination of two vectors from the sum. J. Combin. Theory Ser. B **6**(4), 402–407 (1969)
17. Lindström, B.: On $B_2$-sequences of vectors. J. Number Theory **4**, 261–265 (1972)

18. Raz, R., Yehudayoff, A.: Multilinear formulas, maximal-partition discrepancy and mixed-sources extractors. J. Comput. Syst. Sci. **77**(1), 167–190 (2011)
19. Schnorr, C.P.: A lower bound on the number of additions in monotone computations. Theor. Comput. Sci. **2**(3), 305–315 (1976)
20. Sidon, S.: Ein Satz über trigonometrische Polynome und seine Anwendung in der Theorie der Fourier-Reihen. Math. Ann. **106**(1), 536–539 (1932)
21. Srinivasan, S.: Strongly exponential separation between monotone VP and monotone VNP. ACM Trans. Comput. Theory **12**(4), 23:1–23:12 (2020)
22. Wegener, I.: The complexity of Boolean functions. Wiley-Teubner, Stuttgart (1987)

# Chapter 3
# Rectangle Bounds

**Abstract** Many lower bounds on the Minkowski circuit complexity of sets $A \subseteq \mathbb{N}^n$ of vectors are based on the observation that sets $A$ of small Minkowski circuit complexity can be decomposed into a small number of "balanced" (not necessarily disjoint) rectangles $X + Y \subseteq A$, with various meanings of a rectangle $X + Y$ being "balanced." Thus, in order to show a high lower bound on the Minkowski circuit complexity of a given set $A$, it is enough to show that any such decomposition requires many rectangles. We first prove some general decomposition lemmas and then demonstrate them "in action" by proving lower bounds on the size of tropical circuits solving several basic optimization problems, as well as by proving that Minkowski circuits can be exponentially stronger than monotone arithmetic circuits.

## 3.1 Balanced Decompositions

Many lower bounds on the minimum size $\texttt{Arith}(f)$ of monotone arithmetic $(+, \times)$ circuits computing explicit polynomials $f$ are based on the following structural property of such circuits: if $\texttt{Arith}(f)$ is "small," then the polynomial $f$ can be written as a sum of a "small" number of product polynomials $gh$, where the polynomials $g$ and $h$ are "balanced" in some sense. This way, proving lower bounds on $\texttt{Arith}(f)$ boils down to showing that no balanced product polynomial $gh$ in such a decomposition of the polynomial $f$ can have too many monomials. Since in different contexts, different measures of "balance" are used, let us describe (in Lemmas 3.1 and 3.2) the general idea.

A *component* of a (arithmetic) polynomial $f$ is a polynomial $g$ such that $f = gp + q$ for some polynomials $p$ and $q$. The reason to only consider components of a given polynomial $f$ (instead of arbitrary polynomials) is twofold. First, if $\Phi$ is an arithmetic $(+, \times)$ circuit computing $f$, then the polynomials computed at the gates of $\Phi$ are components of $f$, and we are only interested in these polynomials. Second, knowing the structure of the given polynomial $f$, we can consider measures of polynomials that are "tailor made" for the components of $f$ (and can even make no sense for other polynomials).

© The Author(s), under exclusive license to Springer Nature Switzerland AG 2023      53
S. Jukna, *Tropical Circuit Complexity*, SpringerBriefs in Mathematics,
https://doi.org/10.1007/978-3-031-42354-3_3

A measure $\mu$ on components of a polynomial $f$ assigns nonnegative numbers $\mu(g)$ to components $g$ of $f$. Such a measure is a *norm measure* if it is *normalized* in that $\mu(g) \leqslant 1$ holds for constant polynomials $g = c \in \mathbb{R}_+$ and polynomials $g = x_i$ consisting of just one variable and is *subadditive* in that $\mu(g + h) \leqslant \mu(g) + \mu(h)$ and $\mu(gh) \leqslant \mu(g) + \mu(h)$ hold for any two components $g$ and $h$ of $f$. For example, $\mu(g)$ = degree of $g$, and $\mu(g)$ = a number of variables on which $g$ depends are norm measures.

The following lemma holds for any polynomial $f(x_1, \ldots, x_n)$, any subadditive norm measure $\mu$ of components of $f$ with $\text{nor} f \geq 1$, and any threshold $r \geq 1$.

**Lemma 3.1** *If* $\texttt{Arith}(f) \leqslant t$, *then* $f$ *can be written as a sum* $f = f_0 + \sum_{i=1}^t g_i h_i$, *where* $\mu(f_0) \leqslant r$ *and* $r < \mu(g_i) \leqslant 2r$ *for all* $i$.

**Proof** If $\mu(f) \leqslant r$, then there is nothing to prove (we can take $f_0 := f$ then). So, suppose that $\mu(f) > r$, and let $\Phi$ be a monotone arithmetic circuit of size $t = \texttt{Arith}(f)$ computing the polynomial $f$. For a gate $v$ of $\Phi$, let $f_v$ be the polynomial computed at $v$. By Corollary 1.13, the polynomial $f_v$ is also (syntactically) produced at that gate.

Let $o$ be the output gate of $\Phi$; hence, $\mu(f_o) = \mu(f) > r$. Start at the gate $o$, and traverse the circuit backward by always choosing the child $v$ with the larger norm $\mu(f_v)$. Since $\mu(f_o) > r$ and since the norm of polynomials computed at the input gates is $\leqslant 1 \leqslant r$, we will reach a gate $v$ such that $\mu(f_v) > r$, but $\mu(f_u) \leqslant r$ and $\mu(f_w) \leqslant r$ hold for the gates $u$ and $w$ entering $v$. By the subadditivity of $\mu$, we have $\mu(f_v) \leqslant \mu(f_u) + \mu(f_w) \leqslant 2r$. Hence, $r < \mu(f_v) \leqslant 2r$.

Let $\Psi(x_1, \ldots, x_n, y)$ be the circuit obtained from our circuit $\Phi(x_1, \ldots, x_n)$ by replacing the gate $v$ with a new input variable $y$ and removing the two edges entering $v$. Let $P(x_1, \ldots, x_n, y)$ be the polynomial produced by the new circuit $\Psi$. Hence, our polynomial $f$ produced by the original circuit $\Phi$ is $f(x_1, \ldots, x_n) = P(x_1, \ldots, x_n, f_v)$. We can write the polynomial $P$ as $P = y \cdot p + q$, where $p = p(x_1, \ldots, x_n, y)$ is some polynomial, and $q = P(x_1, \ldots, x_n, 0)$ is the polynomial produced by the circuit $\Psi_{y=0}$ obtained from $\Psi$ after substitution $y = 0$; note that the polynomial $y \cdot p = y \cdot p(x_1, \ldots, x_n, y)$ consists of all terms of $P$ containing the variable $y$ (with some nonzero degrees), and $q$ consists of all remaining terms of $P$. Thus, the polynomial $f$ produced by the original circuit $\Phi$ was of the form $f = g_1 \cdot h_1 + q$, where $g_1 = f_v$, and $h_1 = p(x_1, \ldots, x_n, f_v)$ is the polynomial obtained from $p$ after substitution $y = f_v$. The circuit $\Psi_{y=0}$ has size $t - 1$ and produces the polynomial $q$. If $\mu(q) \leqslant r$, then we are done. Otherwise, we can apply the same argument to the circuit $\Psi_{y=0}$ producing the polynomial $q$. Since after each application, the number of gates decreases by at least one, we will end up with the desired decomposition of the original polynomial $f$.                                          □

Lemma 3.1 gives a similar decomposition of sets of vectors produced by Minkowski $(\cup, +)$ circuits. A *component* of a set $A \subseteq \mathbb{N}^n$ of vectors is a set $X \subseteq \mathbb{N}^n$ of vectors such that $X + Y \subseteq A$ holds for some set $Y \subseteq \mathbb{N}^n$. Note that every subset $X \subseteq A$ of $A$ is also a component of $A$ because then $X + \{\vec{0}\} \subseteq A$ holds. A *measure* on the components of a finite set $A \subseteq \mathbb{N}^n$ of vectors assigns nonnegative numbers $\mu(X)$ to the components $X$ of $A$. Such a measure is a *norm measure* if it is *normalized* in that $\mu(\{x\}) \leqslant 1$ holds for $x \in \{\vec{0}, \vec{e}_1, \dots, \vec{e}_n\}$ and is *subadditive* in that $\mu(X \cup Y) \leqslant \mu(X) + \mu(Y)$ and $\mu(X + Y) \leqslant \mu(X) + \mu(Y)$ hold for any two components $X$ and $Y$ of $A$.

The following lemma holds for any finite set $A \subseteq \mathbb{N}^n$, any norm measure $\mu$ on the components of $A$ with $\mu(A) \geq 1$, and any threshold $r \geq 1$.

**Lemma 3.2** *If* $\text{L}(A) \leqslant t$, *then* $A$ *can be written as a union* $A = A_0 \cup A_1 \cup \cdots \cup A_t$, *where* $\mu(A_0) \leqslant r$ *and each* $A_i = X_i + Y_i$ *for* $i = 1, \dots, t$, *is a rectangle satisfying* $r < \mu(X_i) \leqslant 2r$.

**Proof** By Lemma 1.14, $\text{Arith}(f) \leqslant \text{L}(A)$ holds for some polynomial $f(x) = \sum_{a \in A} c_a \prod_{i=1}^{n} x_i^{a_i}$ with $A$ as its set of exponent vectors and some positive integer coefficients $c_a > 0$. Hence, we also have $\text{Arith}(f) \leqslant t$ for this polynomial. For a polynomial $g$, let $E_g \subseteq \mathbb{N}^n$ denote its set of exponent vectors. To apply Lemma 3.1, we will use the norm measure $\mu$ on the components of the set $A = E_f$ to define a norm measure $\nu$ on the components $g$ of the polynomial $f$. If $g$ is a component of $f$, then $f = gp + q$ holds for some polynomials $p$ and $q$. Since then $E_f = (E_g + E_p) \cup E_q$, the set $E_g$ is a component of the set $A = E_f$. So, we can define the norm measure $\nu(g)$ of components $g$ of the polynomial $f$ by $\nu(g) := \mu(E_g)$. Since the measure $\mu$ of components of $A$ is normalized and subadditive, the measure $\nu$ of components of the polynomial $f$ is also normalized and subadditive. By Lemma 3.1, the polynomial $f$ can be written as a sum $f = f_0 + \sum_{i=1}^{t} g_i h_i$ of products of polynomials, where $\nu(f_0) \leqslant r$ and $r < \nu(g_i) \leqslant 2r$ for all $i$. By taking $A_0 = E_{f_0}$, $X_i = E_{g_i}$, and $Y_i = E_{h_i}$, we have a desired decomposition $A = A_0 \cup (X_1 + Y_1) \cup \cdots \cup (X_t + Y_t)$ of our set $A$ with $\mu(A_0) \leqslant r$ and $r < \mu(X_i) \leqslant 2r$ for all $i$. $\square$

By taking specific norm measures, various versions of Lemma 3.2 can be obtained.

Recall that the *degree* of a vector $a \in \mathbb{N}^n$ is the sum $|a| = a_1 + \cdots + a_n$ of its entries. A set of vectors is *homogeneous* if all its vectors have the same degree. A rectangle $X + Y$ is *m-balanced* if the set $X$ is homogeneous of degree $r$ lying between $m/3$ and $2m/3$.

**Lemma 3.3 (Hyafil [9], Valiant [18])** *Let* $A \subseteq \mathbb{N}^n$ *be homogeneous of degree* $m \geqslant 3$. *Then* $A$ *can be written as a union of at most* $\text{L}(A)$ *m-balanced rectangles.*

**Proof** Let $t = \text{L}(A)$. Since the set $A$ is homogeneous, every component $X$ of $A$ must also be homogeneous. We are going to apply Lemma 3.2 with $r := m/3$ and use the measure $\mu(X) := \deg(X)$, where $\deg(X)$ is the degree of vectors $x \in X$. This measure is clearly normalized. Since $\deg(X + Y) = \deg(X) + \deg(Y)$ and $\deg(X \cup Y) = \max\{\deg(X), \deg(Y)\} \leqslant \deg(X) + \deg(Y)$, the measure is also subadditive. So, by Lemma 3.2, we can write the set $A$ as the union $A = A_0 \cup A_1 \cup \cdots \cup A_t$, where $\deg(A_0) \leqslant m/3$ and each set $A_i$ for $i = 1, \dots, t$ is a rectangle $A_i = X_i + Y_i$ satisfying $m/3 \leqslant \deg(X_i) \leqslant 2m/3$. Since $m/3 < m = \deg(A)$, the set $A_0$ in this decomposition must be empty. $\square$

To give an another application of Lemma 3.2 (which we will use in Sect. 3.3), recall that the *support* of a vector $x \in \mathbb{N}^n$ is the set $\sup(x) = \{i : x_i \neq 0\}$ of its nonzero positions. The *support* of a set $X \subseteq \mathbb{N}^n$ of vectors is the union

$$\sup(X) := \bigcup_{x \in X} \sup(x)$$

of supports of its vectors. Thus, $\sup(X)$ is the set of all positions $i \in [n]$ in which at least one vector of $X$ has a nonzero value. Given a subset $W \subseteq [n]$ of positions, we say that a rectangle $X + Y$ is *W-balanced* if $|\sup(X) \cap W| \leqslant 2|W|/3$ and $|\sup(Y) \cap W| \leqslant 2|W|/3$. In particular, a rectangle $X + Y$ is *[n]-balanced* if $|\sup(X)| \leqslant 2n/3$ and $|\sup(Y)| \leqslant 2n/3$.

**Lemma 3.4 (Raz and Yehudayoff [15], Korhonen [12])** *Let* $A \subseteq \{0, 1\}^n$. *If* $L(A) \leqslant t$, *then for every* $W \subseteq [n]$ *with* $|W| \geq 3$ *the set* $A$ *can be written as a union of at most* $t + 1$ *W-balanced rectangles.*

**Proof** We are going to apply Lemma 3.2 with $r := |W|/3$ and with the measure $\mu(X) := |\sup(X) \cap W|$. This measure is clearly normalized. Since both $\sup(X \cup Y)$ and $\sup(X + Y)$ are equal to $\sup(X) \cup \sup(Y)$, the measure is also subadditive. So, by Lemma 3.2, we can write the set $A$ as the union $A = A_0 \cup A_1 \cup \cdots \cup A_t$, where $|\sup(A_0) \cap W| \leqslant |W|/3$ and each $A_i = X_i + Y_i$ for $i = 1, \ldots, t$ is a rectangle satisfying $|W|/3 \leqslant |\sup(X_i) \cap W| \leqslant 2|W|/3$. Since each rectangle $X_i + Y_i \subseteq A$ must consist of 0-1 vectors (of the set $A$), we also have $\sup(X_i) \cap \sup(Y_i) = \emptyset$. Together with $|\sup(X_i) \cap W| \geqslant |W|/3$, this yields $|\sup(Y_i) \cap W| \leqslant |W| - |\sup(X_i)| \leqslant 2|W|/3$ for each $i = 1, \ldots, t$. It remains to write the first set $A_0$ as the rectangle $A_0 = X_0 + Y_0$ with $X_0 = A_0$ and $Y_0 = \{\vec{0}\}$. $\qquad\square$

We now turn to applications of these general decomposition lemmas.

## 3.2 Matchings, Paths, and Walks

For an integer $r \geqslant 1$ and a set $A \subseteq \{0, 1\}^n$ of vectors, let $\#_r(A)$ be the maximum number of vectors $a \in A$ sharing $r$ or more 1s in common. That is, $\#_r(A)$ is the maximum of $|\{a \in A : a \geqslant b\}|$ over all vectors $b \in \{0, 1\}^n$ with exactly $r$ ones. Note that $\#_r(A)$ is decreasing as a function of $r$: $|A| = \#_0(A) \geqslant \#_1(A) \geqslant \ldots \geqslant \#_d(A) = 1$, where $d$ is the maximum degree of a vector in $A$.

**Lemma 3.5** *Let* $m \geqslant 3$ *and let* $A \subseteq \{0, 1\}^n$ *be a homogeneous set of degree* $m \geqslant 3$. *Then there is an* $r$ *between* $m/3$ *and* $2m/3$ *such that*

$$L(A) \geqslant \frac{|A|}{\#_r(A) \cdot \#_{m-r}(A)}. \tag{3.1}$$

***Proof*** By Lemma 3.3, there are at most $t = \text{L}(A)$ $m$-balanced rectangles $X+Y \subseteq A$ such that every vector of $A$ belongs to at least one of them. Take any such rectangle $X + Y \subseteq A$ and any two vectors $x \in X$ and $y \in Y$. Since $x + y$ must be a 0-1 vector, we have $\langle x, y \rangle = 0$. Since the rectangle is $m$-balanced, we have $|x| = r$ for some $m/3 \leqslant r \leqslant 2m/3$, and since $\langle x, y \rangle = 0$, we also have $|y| \geqslant m - |x| = m - r$. Since all vectors of the translate $x + Y = \{x + y : y \in Y\}$ contain the vector $x$, the inclusion $x + Y \subseteq A$ yields $|Y| = |x + Y| \leqslant \#_{|x|}(A) = \#_r(A)$. Similarly, since all vectors of the translate $X + y$ contain the vector $y$, the inclusion $X + y \subseteq A$ yields $|X| = |X + y| \leqslant \#_{|y|}(A) \leqslant \#_{m-r}(A)$. Thus, for each of these $t$ rectangles $X + Y \subseteq A$, there is an $r$ between $m/3$ and $2m/3$ such that $|X + Y| \leqslant \#_r(A) \cdot \#_{m-r}(A)$. By taking such an $r$ with largest $\#_r(A) \cdot \#_{m-r}(A)$, the lower bound (3.1) on the number $t$ of rectangles follows.                                                                     $\square$

Recall that a *walk* of length $l$ between two nodes $s$ and $t$ is a nonempty alternating sequence $v_0 e_0 v_1 e_1 \ldots e_{l-1} v_l$ of nodes and edges, where $v_0 = s$, $v_l = t$, and $e_j = \{v_j, v_{j+1}\}$ for all $j < l$; note that one and the same edge $e_j$ can appear many times. Such a walk is a (simple) *path* if all vertices $v_0, v_1, \ldots, v_l$ are distinct. A *Hamiltonian path* is a path through all nodes of the underlying graph. View graphs as sets of their edges, fix an ordering of all $m := \binom{n}{2}$ edges of the complete graph $K_n$ on $\{1, \ldots, n\}$, as well as an ordering of all $n^2$ edges of a complete bipartite $n \times n$ graph $K_{n,n}$, and consider the following sets of feasible solutions:

o  $P_n \subseteq \{0, 1\}^m$: characteristic vectors of paths from 1 to $n$ in $K_n$.
o  $H_n \subseteq P_n$: characteristic vectors of Hamiltonian paths from 1 to $n$ in $K_n$.
o  $M_n \subseteq \{0, 1\}^{n^2}$: characteristic vectors of perfect matchings in $K_{n,n}$.
o  $W_n \subseteq \mathbb{N}^m$: the set of vectors $b \in \mathbb{N}^n$ corresponding to all walks in $K_n$ of length at most $n - 1$ between the nodes 1 and $n$. That is, we associate with each such walk the vector $(b_1, \ldots, b_m) \in \mathbb{N}^m$, where $b_i \in \mathbb{N}$ is the number of times the $i$th edge of $K_n$ (in our fixed order) appears in that walk; hence, $\sum_{i=1}^m b_i \leqslant n - 1$ for all $b \in W_n$;

**Theorem 3.6** *For all sufficiently large $n$, the following hold:*

(1) $\text{Min}(M_n) = \text{Max}(M_n) = \text{L}(M_n) = 2^{\Omega(n)}$.
(2) $\text{Min}(H_n) = \text{Max}(H_n) = \text{L}(H_n) = 2^{\Omega(n)}$.
(3) $\text{Bool}(P_n) = \mathcal{O}(n^3)$ *and* $\text{Min}(P_n) = \text{Min}(W_n) \leqslant \text{L}(W_n) = \mathcal{O}(n^3)$, *but both* $\text{L}(P_n)$ *and* $\text{Max}(P_n)$ *are at least* $2^{\Omega(n)}$.

Using more accurate arguments, Jerrum and Snir [10] have shown a stronger lower bound $\text{Min}(H_n) = \Omega(n^2 2^n)$, which, up to multiplicative constant, already matches the upper bound $\text{Min}(H_n) = \mathcal{O}(n^2 2^n)$ resulting from the Bellman–Held–Karp DP algorithm [2, 7] (see Example 1.9).

***Proof*** To show claim (1), observe that the set $M_n$ is homogeneous (of degree $n$). So, the equalities $\text{Min}(M_n) = \text{Max}(M_n) = \text{L}(M_n)$ follow from Lemma 1.34. It remains to show the lower bound $\text{L}(M_n) = 2^{\Omega(n)}$. Every matching with $r$ edges is contained in $(n - r)!$ perfect matchings. Thus, $\#_r(M_n) \leqslant (n - r)!$, and we obtain

$\#_r(M_n) \cdot \#_{n-r}(M_n) \leqslant (n-r)!r! = n!\binom{n}{r}^{-1}$. Since $\binom{n}{r} \geqslant \binom{n}{n/3}$ for every $r$ between $n/3$ and $2n/3$, Lemma 3.5 gives the desired lower bound $\text{L}(M_n) \geqslant \binom{n}{r} \geqslant \binom{n}{n/3} \geqslant 3^{n/3}$.

To show claim (2), observe that since the set $H_n$ is homogeneous (of degree $n-1$), the equalities $\text{Min}(H_n) = \text{Max}(H_n) = \text{L}(H_n)$ follow (again) from Lemma 1.34. The proof of $\text{L}(H_n) = 2^{\Omega(n)}$ is almost the same as for the perfect matchings: we have $|H_n| = (n-1)!$ Hamiltonian 1-to-$n$ paths, and at most $(n-1-r)!$ of them can contain a fixed set of $r$ edges.

To show claim (3), observe that every path is a walk, and every walk contains a path. So, since the input weights are nonnegative, the minimization problems on both sets $P_n$ and $W_n$ are equivalent. In particular, $\text{Min}(P_n) = \text{Min}(W_n)$ holds. The upper bounds $\text{Min}(W_n) \leqslant \text{L}(W_n) = \mathcal{O}(n^3)$ and $\text{Bool}(P_n) = \mathcal{O}(n^3)$ follow by implementing the Bellman–Ford–Moore shortest path DP algorithm as a tropical $(\text{min}, +)$ or Boolean $(\vee, \wedge)$ circuit (see Example 1.7) because $W_n$ is the set of "exponent" vectors produced by that circuit. To show the lower bounds, observe that vectors of the higher envelope $\lceil P_n \rceil$ of $P_n$ correspond to Hamiltonian paths from vertex 1 to vertex $n$. We thus have $\lceil P_n \rceil = H_n$. Together with Lemma 1.32, the lower bound (2) yields $\text{L}(P_n) \geqslant \text{L}(\lceil P_n \rceil) = \text{L}(H_n) = 2^{\Omega(n)}$. Together with Lemma 1.34, we also have $\text{Max}(P_n) \geqslant \text{L}(\lceil P_n \rceil) = 2^{\Omega(n)}$.                            □

Theorem 3.6 gives a direct and simple application of the decomposition lemmas. In the next four sections, we will see that some applications may already require nontrivial combinatorial or probabilistic tools.

## 3.3  Independent Sets

Let $G = (V, E)$ be an undirected graph on $V = \{1, \ldots, n\}$. An *independent set* in $G$ is a set $I \subseteq V$ of vertices without any edge of $G$ lying between them. The *maximum weight independent set* problem (or the *MIS problem*) on $G$ is, given an assignment $x : V \to \mathbb{R}_+$ of nonnegative weights to the vertices of $G$, to compute

$$\text{MIS}(x) = \max_I \sum_{v \in I} x(v),$$

where the maximum is over all independent sets $I \subseteq V$ in $G$. We already know (Example 1.10) that the MIS problem on $n$-vertex trees can be solved by a $(\text{max}, +)$ circuit of size $\mathcal{O}(n)$. But what about the MIS problem on other graphs $G$?

The *treewidth* of a graph is an important graph parameter. Roughly, the treewidth of a graph captures how similar is a graph to a tree. In the proof of the following theorem, familiarity with the concept of treewidth is not assumed: we will only need one structural property of graphs of large treewidth established by Robertson and Seymour [16] (Lemma 3.10 below).

But for an interested reader, let us recall the standard definition of this concept. A *tree decomposition* of a graph $G$ is a tree $T$, whose nodes (called *bags*) are subsets of vertices of $G$, and which satisfies three conditions: (1) every vertex of $G$ must belong to at least one bag, (2) both endpoints of each edge of $G$ belong to at least one bag, and (3) for every vertex $i$ of $G$, the bags containing $i$ must form a subtree. The *treewidth* of $G$ is the minimum, over all tree decompositions $T$, of the maximum size of a bag in $T$ minus one.

Treewidth also has an equivalent definition via so-called brambles. A *bramble* in a graph $G = (V, E)$ is a family $\mathcal{F} \subseteq 2^V$ of subsets of its vertices such that for any two (not necessarily distinct) sets $S, T \in \mathcal{F}$, the induced subgraph $G[S \cup T]$ is connected. That is, a bramble is a family $\mathcal{F}$ of connected subgraphs of $G$ such that *any* two of these subgraphs have a nonempty intersection or are joined by an edge. The *order* of a bramble $\mathcal{F}$ is the least number of vertices required to intersect all sets of $\mathcal{F}$. The *bramble number* of $G$ is the maximum order of all brambles of $G$. Seymour and Thomas [17] proved that a graph has treewidth $w$ iff it has bramble number $w + 1$. Thus, to show that a given graph $G$ has treewidth $\geqslant w$, it is enough to exhibit a bramble $\mathcal{F}$ in $G$ such that every set intersecting all sets of $\mathcal{F}$ must have $\geqslant w + 1$ vertices. It is folklore that an $n \times n$ grid $G = (V, E)$ has treewidth $n$.

A bramble $\mathcal{F}$ of order $n + 1$ in $G$ can be constructed as follows: we have one set that contains all vertices on the bottom row, one set that contains all vertices on the last column except the last one, and $(n - 1)^2$ crosses, each consisting of the first $n - 1$ vertices of one of the first $n - 1$ rows and the first $n - 1$ vertices of one of the first $n - 1$ columns. A set $S \subseteq V$ that intersects all sets of the bramble $\mathcal{F}$ must contain at least $n - 1$ vertices to intersect the crosses (if $|S| \leqslant n - 2$, then $S$ must miss at least one row and at least one column of the $(n - 1) \times (n - 1)$ subgrid) and one vertex in each of the remaining two sets.

**Theorem 3.7 (Korhonen [12])** *Let $G$ be a graph with maximum degree $\leqslant d$ and treewidth $\geqslant w$. Then every (max, +) circuit solving the MIS problem on $G$ must have $2^{\Omega(w/d)}$ gates.*

In particular, since the $n \times n$ grid $G$ has maximum degree $d = 4$ and treewidth $n$, we have the following lower bound.

**Corollary 3.8** *Every (max, +) circuit solving the MIS problem on the $n \times n$ grid must have $2^{\Omega(n)}$ gates.*

**Remark 3.9** It is shown in [12] that Theorem 3.7 is optimal up to a factor $d$ in the sense that for each pair $w, d$ one can construct a graph $G$ with treewidth $\Omega(w)$ and maximum degree $\mathcal{O}(d)$ such that the MIS problem on $G$ can be solved by a (max, +) formula of size $d2^{w/d}$.                                                                                     □

Crucial in the entire proof of Theorem 3.7 will be the following combinatorial property of graphs of large treewidth. A *separation* of a graph $G = (V, E)$ is an ordered triple of vertex sets $(P, S, Q)$ such that $P, S,$ and $Q$ are disjoint, $P \cup S \cup Q = V$, and no vertex of $P$ is adjacent to a vertex of $Q$, that is, $E \cap (P \times Q) = \emptyset$. Thus, if we delete all vertices of $S$, then each connected component of the remaining graph lies entirely in $P$ or in $Q$. The *order* of a separation $(P, S, Q)$ is $|S|$.

Given a set $W \subseteq V$ of vertices, say that a separation $(P, S, Q)$ is $W$-*balanced* if $|P \cap W| \leqslant 2|W|/3$ and $|Q \cap W| \leqslant 2|W|/3$. In particular, if $W = V$, then every partition $V = P \cup S \cup Q$ with $|P| \leqslant 2|V|/3$ and $|Q| \leqslant 2|V|/3$ is $W$-balanced.

Robertson and Seymour [16] have proved that if for every $W \subseteq V$ the graph $G$ has a $W$-balanced separation of order $\leqslant k$, then the treewidth of $G$ is $< 4k$. In other words, the order of balanced separations in graphs of large treewidth is large:

**Lemma 3.10 (Robertson and Seymour [16])** *If a graph* $G = (V, E)$ *has treewidth at least 4k, then there is a vertex set* $W \subseteq V$ *such that* $|S| \geqslant k$ *holds for any $W$-balanced separation* $(P, S, Q)$ *of $G$.*

Theorem 3.7 is a direct consequence of the following two lemmas (Lemmas 3.11 and 3.12). We say that a set $S$ *avoids* a set $T$ if $S \cap T = \emptyset$.

**Lemma 3.11** *Let* $G = (V, E)$ *be an n-vertex graph with maximum degree at most $d$ and treewidth* $w \geqslant 4k$. *If the MIS problem on $G$ can be solved by a* (max, +) *circuit of size $t$, then there is a family* $\mathcal{F} \subseteq \binom{V}{k}$ *of* $|\mathcal{F}| \leqslant t + 1$ *k-element subsets of vertices such that every independent set of $G$ avoids at least one set of $\mathcal{F}$.*

**Proof** Let $A \subseteq \{0, 1\}^n$ be the set of characteristic vectors of maximal independent sets in $G = (V, E)$, that is, of independent sets contained in no other independent set. Note that the set $A$ is an antichain, and the downward closure $A^{\downarrow} = \{b : b \leqslant a$ for some $a \in A\}$ of $A$ consists of characteristic 0-1 vectors of *all* independent sets in $G$. Recall that the *support* of a vector $x \in \mathbb{N}^n$ is the set $\sup(x)$ of its nonzero positions. In particular, for every vector $x \in A^{\downarrow}$, $\sup(x) \subseteq V$ is an independent set in our graph $G$. The *support* of a set $X \subseteq \mathbb{N}^n$ of vectors is the union $\sup(X) = \bigcup_{x \in X} \sup(x)$ of the supports of its vectors.

Let $t = \text{Max}(A)$. Thus, $t$ is the smallest size of a (max, +) circuit solving the MIS problem on $G$. Since the set $A$ consists of only 0-1 vectors and is an antichain, Lemma 1.31 gives us a set $B$ such that $A \subseteq B \subseteq A^{\downarrow}$ and $\text{L}(B) = t$.

Now let $W \subseteq V$ be the set ensured by Lemma 3.10. Using this vertex set $W$, Lemma 3.4 implies that the set $B$ can be written as a union of at most $t+1$ rectangles $X_i + Y_i \subseteq B$ such that $\sup(X_i) \cap \sup(Y_i) = \emptyset$, $|\sup(X_i) \cap W| \leqslant 2|W|/3$, and $|\sup(Y_i) \cap W| \leqslant 2|W|/3$. Each of these rectangles $X_i + Y_i$ gives us a partition of vertices of $G$ into three disjoint sets

$$P_i := \sup(X_i) \qquad Q_i := \sup(Y_i) \qquad S_i := V \setminus \sup(X_i + Y_i).$$

Since $A \subseteq B$, the characteristic 0-1 vector $a_I \in \{0, 1\}^n$ of every maximal independent set $I \subseteq V$ of $G$ must belong to at least one of the rectangles $X_i + Y_i$. In particular, every independent set $I$ of $G$ must lie in at least one of the sets $\sup(X_i + Y_i) = P_i \cup Q_i$. On the other hand, since $B \subseteq A^{\downarrow}$ and $X_i + Y_i \subseteq B$, every vector $x + y$ with $x \in X_i$ and $y \in Y_i$ is contained in at least one vector of $A$, that is, $x + y$ is the characteristic vector of an independent set of $G$. This means that no vertex $u \in \sup(x)$ can be adjacent to any vertex $v \in \sup(y)$. Since this holds for all vectors $x \in X_i$ and all vectors $y \in Y_i$, no vertex of $P_i = \sup(X_i)$ can be adjacent to a vertex of $Q_i = \sup(Y_i)$. So, each of $t + 1$ rectangles $X_i + Y_i$ gives us a $W$-balanced separation $(P_i, S_i, Q_i)$ of $G$. Since the graph $G$ has treewidth $w \geqslant 4k$, Lemma 3.10 ensures that $|S_i| \geqslant w/4 \geqslant k$ holds for all $i = 1, \ldots, t + 1$. Since the sets $S_i$ and $P_i \cup Q_i$ are disjoint, no independent set $I \subseteq V$ can intersect *all* sets $S_i = V \setminus (P_i \cup Q_i)$, $i = 1, \ldots, t + 1$, for, otherwise, the independent set $I$ would lie in none of the sets $\sup(X_i + Y_i)$.                                                                    $\square$

**Lemma 3.12** *Let $G = (V, E)$ be a graph with maximum degree $d \geqslant 2$, and let $\mathcal{F} \subseteq \binom{V}{k}$ be some family of $k$-element sets of its vertices. If every independent set of $G$ avoids at least one member of $\mathcal{F}$, then $|\mathcal{F}| = 2^{\Omega(k/d)}$.*

That is, given any collection $S_1, \ldots, S_t \subseteq V$ of $t = 2^{o(k/d)}$ $k$-element subsets of vertices, it is possible to pick (not necessarily distinct) vertices $v_1 \in S_1, \ldots, v_t \in S_t$ such that there are no edges between these vertices.

To prove Lemma 3.12, Korhonen used the lopsided Lovász Local Lemma [1, 6]. Let $A_1, \ldots, A_m$ be events in a probability space. Think of them as some "bad" events, and our goal is to show that $\Pr\{\bigcap_{i=1}^{m} \overline{A_i}\} > 0$, i.e., that *none* of them happens with a nonzero probability. If these events were independent, then we would have no problem: then $\Pr\{\bigcap_{i=1}^{m} \overline{A_i}\} = \prod_{i=1}^{m} \Pr\{\overline{A_i}\}$ would hold. In the case when some dependencies between the events do exist, Lovász Local Lemma comes to rescue.

Let $\Gamma$ be an undirected graph on the set $[m] = \{1, \ldots, m\}$ of indices $i$ of the events $A_i$. For a vertex $i \in [m]$, let $N(i) \subseteq [m]$ denote the set of all vertices adjacent to $i$ in $\Gamma$. The graph $\Gamma$ is a *negative dependency graph* of the events $A_1, \ldots, A_m$ if $\Pr\{A_i \cap \bigcup_{j \in J} A_j\} \geqslant \Pr\{A_i\} \cdot \Pr\{\bigcup_{j \in J} A_j\}$ holds for every node $i \in [m]$ and every subset $J \subseteq [m] \setminus N(i)$. In particular, $\Pr\, A_i \cap A_j \geq \Pr\, A_i \cdot \Pr\, A_j$ holds for any two non-adjacent vertices $i$ and $j$. In words, the edges of the negative dependency graph capture all negative correlations between the events.[1] The lopsided Lovász Local Lemma states [1]:

- $\Pr\{\bigcap_{i=1}^{m} \overline{A_i}\} > 0$ holds if there exist real numbers $x_1, \ldots, x_m$ with $0 < x_i < 1$ such that $\Pr\{A_i\} \leqslant x_i \prod_{j \in N(i)} (1 - x_j)$ for all $i \in [m]$.

*Proof of Lemma 3.12* It is enough to show that if $5|\mathcal{F}| + 1 \leqslant e^{k/(6d)}$, then some independent set of $G$ intersects all sets in $\mathcal{F}$. To show the existence of such an independent set, we use the Local Lemma. Let $I \subseteq V$ be a random set of vertices of our graph $G$ with each vertex included randomly and independent with probability $p := 1/(2d)$. Our bad events are $A_e$ for each edge $e$ indicating that both endpoints of $e$ are selected in $I$ and $A_S$ for each $S \in \mathcal{F}$ indicating that $S \cap I = \emptyset$. Hence, $\Pr\{A_e\} = p^2$ and $\Pr\{A_S\} = (1 - p)^{|S|}$. Our goal is to show that, with *positive* probability, *none* of these bad events will happen.

---

[1] Events $A$ and $B$ are negatively correlated if $\Pr\{A \cap B\} < \Pr\{A\} \cdot \Pr\{B\}$, which is equivalent to $\Pr\{A|B\} < \Pr\{A\}$. In particular, independent events are *not* negatively correlated.

It is easy to verify that the edge events $A_e$ have nonnegative correlation with each other and the vertex set events $A_S$ also have nonnegative correlation with each other: if $e, f \in E$, then $\Pr\{A_e \cap A_f\} = \Pr\{e \subseteq I \text{ and } f \subseteq I\} = p^{|e \cup f|} \geq p^{|e|+|f|} = \Pr\{A_e\} \cdot \Pr\{A_f\}$, and if $S, T \in \mathcal{F}$, then $\Pr\{A_S \cap A_T\} = \Pr\{(S \cup T) \cap I = \emptyset\} = (1-p)^{|S \cup T|} \geq (1-p)^{|S|+|T|} = \Pr\{A_S\} \cdot \Pr\{A_T\}$. If $e \cap S = \emptyset$ holds for an edge $e \in E$ and a set $S \in \mathcal{F}$, then $\Pr\{A_e \cap A_S\} = p^2(1-p)^{|S|} = \Pr\{A_e\} \cdot \Pr\{A_S\}$, meaning that there is no correlation between the events $A_e$ and $A_S$ in this case.

Thus, the negative dependency graph $\Gamma$ is a bipartite graph connecting events $A_e$ and $A_S$ iff $e \cap S \neq \emptyset$. Note that each "edge-vertex" $e$ of the graph $\Gamma$ has at most $|\mathcal{F}|$ neighbors, and each "set-vertex" $S$ has at most $d|S|$ neighbors in the graph $\Gamma$.

For all edge events $A_e$, we choose $x_e = x := 1/(3d^2 + 1)$, and for all vertex set events $A_S$, we choose $x_S = y := 1/(5|\mathcal{F}| + 1)$. By the lopsided Lovász Local Lemma, it suffices to verify that

$$\Pr\{A_e\} = p^2 \leq x(1-y)^{|\mathcal{F}|} \qquad (3.2)$$

and

$$\Pr\{A_S\} = (1-p)^{|S|} \leq y(1-x)^{d|S|} \qquad (3.3)$$

hold whenever $6|\mathcal{F}| \leq e^{|S|/6d}$. We will use the estimate $1 - t \geq e^{-t-t^2}$ holding for any $0 < t < 0.6$ (see, e.g., [3, Sect. 1.1]); for $r \geq 2$, it yields $1 - 1/r \geq e^{-1/(r-1)}$.

For (3.2), a lower bound on the right-hand side is $xe^{-1/(5|\mathcal{F}|)} \geq xe^{-1/5} = e^{-1/5}/(3d^2 + 1)$, which can be verified to be greater than $p^2 = 1/(4d^2)$ when $d \geq 2$. For (3.3), using $1 - t \leq e^{-t}$, an upper bound on the left-hand side is $e^{-p|S|} = e^{-|S|/(2d)}$, and a lower bound on the right-hand side is $ye^{-d|S|/(3d^2)} = ye^{-|S|/(3d)}$, implying that (3.3) holds if $e^{-|S|/(2d)}e^{|S|/(3d)} \leq y$ or, equivalently, if $e^{-|S|/(6d)} \leq y$. By the choice of $y = 1/(5|\mathcal{F}| + 1)$, this simplifies to $e^{|S|/(6d)} \geq 5|\mathcal{F}| + 1$, which holds by our assumption about the size $|\mathcal{F}|$ of our family $\mathcal{F}$.                    $\square$

## 3.4    Arborescences

Let $\vec{K}_n$ be a directed complete graph on $\{1, \ldots, n\}$. An *arborescence* (or a *directed spanning tree*) in $\vec{K}_n$ is a directed tree $T$ on $[n]$ such that the vertex $r = 1$ (the root) is reachable from every other vertex. Thus, every vertex other than the root has outdegree 1:

Let $\vec{\mathcal{J}}_n$ be the family of all arborescences in $\vec{K}_n$. We view arborescences as sets of their (directed) edges. The *directed spanning tree polynomial* (or the *arborescence polynomial*) is the following homogeneous, multilinear polynomial of degree $n - 1$:

$$\mathrm{DST}_n(x) = \sum_{T \in \vec{\mathcal{J}}_n} \prod_{e \in T} x_e.$$

As before, for a polynomial $f$ with positive coefficients, $\mathrm{Arith}(f)$ denotes the minimum size of a monotone arithmetic $(+, \times)$ circuit computing $f$.

**Theorem 3.13 (Jerrum and Snir [10])** $\mathrm{Arith}(\mathrm{DST}_n) = 2^{\Omega(n)}$.

To prove Theorem 3.13, we will use the following set-theoretic version of the Hyafil–Valiant decomposition lemma (Lemma 3.3). Say that a family $\mathcal{R}$ of sets is a *rectangle* if it is of the form

$$\mathcal{R} = \mathcal{A} \vee \mathcal{B} := \{A \cup B : A \in \mathcal{A} \text{ and } B \in \mathcal{B}\}$$

for some families $\mathcal{A}$ and $\mathcal{B}$ of sets such that $A \cap B = \emptyset$ holds for all $A \in \mathcal{A}$ and $B \in \mathcal{B}$. A family $\mathcal{F}$ of sets is *m-uniform* if $|F| = m$ holds for all $F \in \mathcal{F}$. A rectangle $\mathcal{R} = \mathcal{A} \vee \mathcal{B} \subseteq \mathcal{F}$ is *m-balanced* if the family $\mathcal{A}$ is $r$-uniform for some $m/3 \leqslant r \leqslant 2m/3$. When applied to the set of characteristic 0-1 vectors of the sets $F \in \mathcal{F}$, Lemma 3.3 directly yields the following decomposition.

**Lemma 3.14 (Set-Version of Lemma 3.3)** *Let* $f(x) = \sum_{F \in \mathcal{F}} \prod_{i \in S} x_i$, *where* $\mathcal{F} \subseteq 2^{[n]}$ *is an m-uniform family for* $m \geqslant 3$. *If* $\mathrm{Arith}(f) \leqslant t$, *then* $\mathcal{F}$ *can be written as a union of at most t m-balanced rectangles.*

*Proof of Theorem 3.13* The corresponding to the polynomial $\mathrm{DST}_n$ family is the family $\vec{\mathcal{J}}_n$ of all $|\vec{\mathcal{J}}_n| = n^{n-2}$ arborescences in $\vec{K}_n$ (Moon [14]), each viewed as the set of its (directed) edges. This family is $m$-uniform for $m := n - 1$.

So, fix an $m$-balanced rectangle $\mathcal{R} \subseteq \vec{\mathcal{J}}_n$. The rectangle is of the form $\mathcal{R} = \mathcal{A} \vee \mathcal{B} = \{A \cup B : A \in \mathcal{A} \text{ and } B \in \mathcal{B}\}$ for two families $\mathcal{A}$ and $\mathcal{B}$ of directed graphs on $\{1, \ldots, n\}$ such that for all graphs $A \in \mathcal{A}$ and $B \in \mathcal{B}$ (viewed as sets of their edges), we have $A \cap B = \emptyset$ (the graphs are edge-disjoint), and $A \cup B$ is an arborescence in $\vec{K}_n$. Since the rectangle $\mathcal{R} = \mathcal{A} \vee \mathcal{B}$ is balanced and each arborescence in $\vec{K}_n$ has $m = n - 1$ edges, we have $m/3 \leqslant |A| \leqslant 2m/3$ and $m/3 \leqslant |B| \leqslant 2m/3$ for every arborescence $A \cup B$ with $A \in \mathcal{A}$ and $B \in \mathcal{B}$. By Lemma 3.3, it is enough to show that $|\mathcal{R}| \leqslant 2^{-\Omega(n)} |\vec{\mathcal{J}}_n|$, i.e., that the rectangle $\mathcal{R}$ can contain at most a $2^{-\Omega(n)}$ portion of all arborescences.

Let $V_A$ denote the set of all vertices $i \in [n]$ such that some edge $(i, j)$ with tail $i$ belongs to at least one $A \in \mathcal{A}$; the set $V_B$ is defined similarly. Since graphs $A \in \mathcal{A}$ and $B \in \mathcal{B}$ are edge-disjoint, and since one vertex can be the tail of only one edge of an arborescence, we have $V_A \cap V_B = \emptyset$, and since the rectangle is balanced, we also have $|V_A| \geqslant m/3$ and $|V_B| \geqslant m/3$.

Let $E = \bigcup_{i=2}^{n} E_i$, where $E_i$ is the set of all edges $e = (i, j)$ whose tail is vertex $i$ and which belong to at least one arborescence of the rectangle $\mathcal{R} = \mathcal{A} \vee \mathcal{B}$. Every set $E_i$ lies in the star $S_i = \{(i, j): j \in [n] \setminus \{i\}\}$. So, $|E_i| \leqslant |S_i| = n - 1$ holds for all $i \in \{2, \ldots, n\}$, and we have an upper bound $|E| \leqslant \sum_{i=2}^{n} |S_i| = (n - 1)^2$. But this trivial bound can be improved by the following observation. Suppose that $i \in V_A$ and $j \in V_B$. Then both edges $(i, j)$ and $(j, i)$ belong to the union $\bigcup_{i=2}^{n} S_i$ of all stars. This union contains all $2|V_A| \cdot |V_B|$ directed edges between $V_A$ and $V_B$ (in both directions). But the edges $(i, j) \in V_A \times V_B$ and $(j, i) \in V_B \times V_A$ cannot both appear in $E$, for if they did, the edge $(i, j)$ would appear in some graph $A \in \mathcal{A}$, the edge $(j, i)$ would appear in some graph $B \in \mathcal{B}$, and $A \cup B$ would not be an arborescence:[2] we would have a 2-cycle $i \to j \to i$. So, out of any pair $(i, j)$ and $(j, i)$ of $2|V_A| \cdot |V_B|$ possible edges, at most one can belong to $E$. Since $m = n - 1$ and $|V_A|, |V_B| \geqslant m/3$, this yields

$$|E| = \sum_{i=2}^{n} |E_i| \leqslant \sum_{i=2}^{n} |S_i| - |V_A| \cdot |V_B| = (n - 1)^2 - |V_A| \cdot |V_B|$$

$$\leqslant (n - 1)^2 - (m/3)^2 = (8/9)(n - 1)^2.$$

Each arborescence in $\vec{K}_n$ is specified by a function $h : \{2, 3, \ldots, n\} \to \{1, 2, \ldots, n\}$ with the property that $\forall i \, \exists k \, h^k(i) = 1$; the arborescence specified by $h$ consists of edges $(2, h(2)), (3, h(3)), \ldots, (n, h(n))$. So, arborescences of the rectangle $\mathcal{R}$ correspond to functions $h$ with the property that $(i, h(i)) \in E_i$ for all $i = 2, \ldots, n$. The number of such functions $h$ is $\prod_{i=2}^{n} |E_i|$. The product is maximized when $|E_i|$ is independent of $i$, that is, when $|E_i| = |E|/(n - 1)$ for all $i$. Thus,

$$|\mathcal{R}| \leqslant \prod_{i=2}^{n} |E_i| \leqslant \left(\frac{|E|}{n - 1}\right)^{n-1} \leqslant \left(\frac{8}{9}\right)^{n-1} n^{n-1} = 2^{-\Omega(n)} |\vec{\mathcal{J}}_n|. \qquad \square$$

**Remark 3.15** The arborescence polynomial $\mathrm{DST}_n$ is multilinear and homogeneous. Thus, together with Lemma 1.34, Theorem 3.13 yields a lower bound $2^{\Omega(n)}$ on the size of tropical circuits solving the minimization as well as the maximization problem on the family $\vec{\mathcal{J}}_n$ of arborescences.

## 3.5   Spanning Trees

As shown by Edmonds [5], the downward closure of the family $\vec{\mathcal{J}}_n$ of arborescences is an intersection of two matroids, meaning that the standard greedy algorithm can *approximate* the minimization problem on $\vec{\mathcal{J}}_n$ within the factor 2. Thus,

---

[2] This is a critical difference from *undirected* graphs where both $(i, j)$ and $(j, i)$ represent the same edge $\{i, j\}$, i.e., there are no heads and tails of edges in the undirected case.

Theorem 3.13 shows that approximating (within factor $r = 2$) greedy algorithms can "beat" exactly solving (within factor $r = 1$) pure DP algorithms. But what about the *same* factor $r = 1$ for both types of algorithms: can then greedy also beat pure DP algorithms? Theorem 3.16 below gives an *affirmative* answer.

A classical optimization problem showing the power of the greedy algorithm is the MST problem (the minimum weight spanning tree problem). The *MST problem* on $K_n$ is, given an assignment of nonnegative weights $x_e$ to the edges $e$ of $K_n$, to compute the minimum weight $\sum_{e \in T} x_e$ of a spanning tree $T$ of $K_n$. Recall that a spanning tree of a connected graph is its connected subgraph which has no cycles and includes all vertices of the graph. Since the family $\mathcal{T}_n$ of spanning trees of $K_n$ is the family of bases of a matroid, the greedy algorithm can efficiently solve the MST problem. For example, the well-known greedy algorithm of Kruskal [13] solves the minimum weight spanning tree problem on any $n$-vertex graph using only $\mathcal{O}(n^2 \log n)$ operations.

The arithmetic version of the MST problem is the *spanning tree polynomial* (also known as the *Kirchhoff polynomial*):

$$\mathrm{ST}_n(x) = \sum_{T \in \mathcal{T}_n} \prod_{e \in T} x_e \,,$$

where $\mathcal{T}_n$ is the family of all $|\mathcal{T}_n| = n^{n-2}$ spanning trees $T$ of $K_n$ (i.e., trees including all $n$ vertices of $K_n$).

**Theorem 3.16 (Jukna and Seiwert [11])** $\mathrm{Arith}(\mathrm{ST}_n) = 2^{\Omega(\sqrt{n})}$.

**Remark 3.17** The spanning tree polynomial $\mathrm{ST}_n$ is multilinear and homogeneous. Thus, together with Lemma 1.34, Theorem 3.16 implies that every (min, +) circuit solving the MST problem on $K_n$ must have $2^{\Omega(\sqrt{n})}$ gates.                           □

In the proof of Theorem 3.16, as well as in the next section, we will use the following consequence of known tail inequalities for the hypergeometric distribution.

**Lemma 3.18 (Hoeffding [8], Chvátal [4])** Let $R \subseteq [n]$ be a fixed set of density $p = |R|/n$. Fix any $0 \leqslant \epsilon \leqslant p$, and let $\mathcal{F} \subseteq \binom{[n]}{k}$ be the family of all $k$-element subsets $S$ of $[n]$ such that $|S \cap R| \leqslant (p - \epsilon)k$. Then $|\mathcal{F}| \leqslant e^{-2\epsilon^2 k} \binom{n}{k}$.

That is, if $R \subseteq [n]$ is large, then a "typical" $k$-element subset $S \subseteq [n]$ must have large intersection with $R$.

*Proof* Let $m := pn = |R|$. In the hypergeometric distribution, we have the set $[n]$ of $n$ balls, $|R| = m$ of which are red and $n - m$ are blue. We draw uniformly at random *without* replacement $k$ times, and let $X$ denote the number of red balls in the sample. The number of possibilities to choose a $k$-element subset $S$ of $[n]$ containing $i$ read and $k - i$ blue balls is $\binom{m}{i}\binom{n-m}{k-i}$. So, $\Pr\{X = i\} = \binom{m}{i}\binom{n-m}{k-i}\binom{n}{k}^{-1}$. A special case of Hoeffding's inequality [8] yields $\Pr\{X \leqslant (p - \epsilon)k\} \leqslant e^{-2\epsilon^2 k}$; see

also[3] Chvátal [4] for a direct proof of this special case. It remains to observe that
$\Pr\{X \leqslant (p - \epsilon)k\} = |\mathcal{F}| \cdot \binom{n}{k}^{-1}$.                                                          □

We now turn to the proof of Theorem 3.16. As in the case of arborescences, we are going to apply the set-version of the Hyafil–Valiant decomposition lemma (Lemma 3.14). Each rectangle $\mathcal{R} = \mathcal{A} \vee \mathcal{B} \subseteq \mathcal{T}_n$ is specified by giving two families $\mathcal{A}$ and $\mathcal{B}$ of forests in the complete graph $K_n$ on $\{1, \ldots, n\}$ such that for all forests $A \in \mathcal{A}$ and $B \in \mathcal{B}$ (viewed as sets of their edges), we have $A \cap B = \emptyset$ (the forests are edge-disjoint), and $A \cup B$ is a spanning tree of $K_n$. The rectangle itself is the family $\mathcal{R} = \mathcal{A} \vee \mathcal{B} := \{A \cup B : A \in \mathcal{A} \text{ and } B \in \mathcal{B}\}$ of all resulting spanning trees. Such a rectangle $\mathcal{R} = \mathcal{A} \vee \mathcal{B}$ is balanced if there is an $r$ between $(n - 1)/3$ and $2(n - 1)/3$ such that $|A| = r$ for all forests $A \in \mathcal{A}$; recall that every spanning tree of a graph on $n$ vertices has $n - 1$ edges. So, every forest $B \in \mathcal{B}$ has $|B| = n - 1 - r \geqslant (n - 1)/3$ edges as well.

In the case of perfect matchings (Theorem 3.6), we knew that, in fact, matchings $A$ and $B$ must be even *vertex-disjoint*. Similarly, in the case of arborescences (Theorem 3.13), we knew that partial arborescences $A$ and $B$ must be "almost" *vertex-disjoint*: they cannot share common tails of edges. But now we only know that forests $A \in \mathcal{F}$ and $B \in \mathcal{B}$ are *edge-disjoint*. To get rid of this "lack of structure," the idea is to concentrate on spanning trees of $K_n$ of a special form.

Let $m$ and $d$ be positive integer parameters satisfying $(d+1)m = n$, $m = \Theta(\sqrt{n})$, and $m \leqslant d/32$; these two parameters will be specified later. A *star* $S_u$ is a tree with one vertex $u$, the *center*, adjacent to all the others, which are the *leaves* of the star. A *d-star* is a star $K_{1,d}$ with $d$ leaves. A *star-factor* is a spanning forest $F$ of $K_n$ consisting of $m$ vertex-disjoint $d$-stars. Every star-factor $F$ is contained in at least one spanning tree $T$ of $K_n$ (in fact, in many of them). Such a tree consists of $md$ edges of $m$ stars of $F$ joined by some additional $m - 1$ edges into a tree. We will concentrate on only spanning trees $T \in \mathcal{T}_n$ such that removal of some $m - 1$ edges from $T$ leaves us with a star-factor.

Let $\mathcal{F}$ be the family of all star-factors of $K_n$. For a rectangle $\mathcal{R} \subseteq \mathcal{T}_n$, let $\mathcal{F}_{\mathcal{R}} \subseteq \mathcal{F}$ denote the family of all star-factors $F$ of $K_n$ contained in at least one spanning tree of $\mathcal{R}$. Every covering $\mathcal{T}_n = \mathcal{R}_1 \cup \mathcal{R}_2 \cup \cdots \cup \mathcal{R}_t$ of the family $\mathcal{T}_n$ of all spanning trees of $K_n$ by $t$ balanced rectangles $\mathcal{R}_i$ gives us a covering $\mathcal{F} = \mathcal{F}_{\mathcal{R}_1} \cup \mathcal{F}_{\mathcal{R}_2} \cup \cdots \cup \mathcal{F}_{\mathcal{R}_t}$ of the family $\mathcal{F}$ of all star-factors by $t$ families $\mathcal{F}_{\mathcal{R}_i}$.

Thus, Theorem 3.16 is a direct consequence of the following lemma.

**Lemma 3.19** *There is an absolute constant* $c > 0$ *such that for every balanced rectangle* $\mathcal{R} \subseteq \mathcal{T}_n$, *we have* $|\mathcal{F}_{\mathcal{R}}| \leqslant |\mathcal{F}| \cdot 2^{-c\sqrt{n}}$.

The rest of this section is devoted to the proof of Lemma 3.19. First, observe that every star-factor $F \in \mathcal{F}$ can be constructed using the following procedure:

- Choose a subset of $m$ centers in $[n]$; $\binom{n}{m}$ possibilities.

---

[3] In both papers [8] and [4], this upper bound is only stated for the event $X \geqslant (p + \epsilon)k$, but using the duality (count blue balls instead of red), the same upper bound holds also for the event $X \leqslant (p - \epsilon)k$.

- Divide the remaining $n - m$ vertices into $m$ blocks of size $d$, and connect all vertices of the $i$th block to the $i$th largest of the chosen centers; there are $\binom{n-m}{d,\ldots,d} = \frac{(n-m)!}{d!^m}$ possibilities to do this.

Since different realizations of this procedure lead to different star-factors (as long as $d > 1$), we have

$$|\mathcal{F}| = \binom{n}{m} \frac{(n-m)!}{d!^m}.$$

Now fix a balanced rectangle $\mathcal{R} = \mathcal{A} \vee \mathcal{B}$ lying in the family $\mathcal{T}_n$ of all spanning trees of $K_n$. Our goal is to prove an upper bound $|\mathcal{F}_{\mathcal{R}}| \leqslant |\mathcal{F}| \cdot 2^{-\Omega(\sqrt{n})}$ on the number of star-factors covered by the rectangle $\mathcal{R}$. A general idea is to show that if the above procedure is required to *only* construct star-factors covered by the rectangle $\mathcal{R}$, then the number of possibilities is substantially lowered.

For this, we will use the following structural restriction on the star-factors covered by our rectangle $\mathcal{R}$. We say that a $d$-star $S_u$ (centered in a vertex $u$) is *covered* by the rectangle $\mathcal{R}$ if it is a star of some star-factor $F \in \mathcal{F}_{\mathcal{R}}$.

**Claim 3.20 (Structural Restriction)** *There is a subset $U_0 \subseteq [n]$ of $|U_0| \geqslant n/16$ vertices, and for each $u \in U_0$, a subset $V_u \subseteq [n] \setminus U_0$ of $|V_u| \geqslant n/16$ vertices such that the following holds for every $d$-star $S_u$ covered by the rectangle $\mathcal{R}$:*

$(*)$  *If $u \in U_0$, then $|S_u \cap (\{u\} \times V_u)| \leqslant m$.*

That is, if the center $u$ of a $d$-star $S_u$ covered by the rectangle $\mathcal{R}$ belongs to the set $U_0$, then at most $m = \mathcal{O}(\sqrt{n})$ out of all $|V_u| \geqslant n/16$ possible candidates for the leaves of $S_u$ can belong to $V_u$. We postpone the proof of Claim 3.20 to the end and first show how the proof of Lemma 3.19 can be finished given this claim.

*Proof of Lemma 3.19 Given Claim 3.20* Let $\mathcal{R} \subseteq \mathcal{T}_n$ be a balanced rectangle. Our goal is to show an upper bound $|\mathcal{F}_{\mathcal{R}}| \leqslant |\mathcal{F}| \cdot 2^{-\Omega(\sqrt{n})}$ on the number of star-factors covered by the rectangle $\mathcal{R}$. For this, split the family $\mathcal{F}_{\mathcal{R}}$ into the family $\mathcal{F}_{\mathcal{R}}^0$ of star-factors $F \in \mathcal{F}_{\mathcal{R}}$ with *no* center in the set $U_0$ and the family $\mathcal{F}_{\mathcal{R}}^1$ of all star-factors $F \in \mathcal{F}_{\mathcal{R}}$ with *at least* one center in $U_0$. The goal is to prove the upper bounds $|\mathcal{F}_{\mathcal{R}}^0| \leqslant |\mathcal{F}| \cdot 2^{-\Omega(m)}$ and $|\mathcal{F}_{\mathcal{R}}^1| \leqslant m|\mathcal{F}| \cdot 2^{-\Omega(d)}$. Since both parameters $m$ and $d$ are $\Theta(\sqrt{n})$, the desired upper bound $|\mathcal{F}_{\mathcal{R}}| = |\mathcal{F}_{\mathcal{R}}^0| + |\mathcal{F}_{\mathcal{R}}^1| \leqslant |\mathcal{F}| \cdot 2^{-\Omega(\sqrt{n})}$ follows.

The upper bound $|\mathcal{F}_{\mathcal{R}}^0| \leqslant |\mathcal{F}| \cdot 2^{-\Omega(m)}$ is easy to show. By Claim 3.20, we know that there are $|U_0| = \Omega(n)$ "critical" vertices. Each star-factor in $\mathcal{F}_{\mathcal{R}}^0$ can be constructed in the same way as above with the difference that centers can only be chosen from $[n] \setminus U_0$, not from the entire set $[n]$. Thus,

$$\frac{|\mathcal{F}_{\mathcal{R}}^0|}{|\mathcal{F}|} \leqslant \binom{n - |U_0|}{m} \cdot \binom{n}{m}^{-1} \leqslant e^{-|U_0| \cdot m/n} = 2^{-\Omega(m)}.$$

Here we used the second of the two inequalities holding for all $b \leqslant b + x < a$:

$$\left(\frac{a-b-x}{a-x}\right)^x \leqslant \binom{a-x}{b}\binom{a}{b}^{-1} \leqslant \left(\frac{a-b}{a}\right)^x . \tag{3.4}$$

These inequalities can be shown by writing the binomial coefficients as quotients of products of factorials (see, e.g., [3, Sect. 1.1]).

Computations in the proof of an upper bound $|\mathcal{F}_{\mathcal{R}}^1| \leqslant m|\mathcal{F}| \cdot 2^{-\Omega(d)}$ are a bit more involved. By Claim 3.20, we have $|V_u| \geqslant n/16$ for each vertex $u \in U_0$. By the definition of $\mathcal{F}_{\mathcal{R}}^1$, every star-factor $F \in \mathcal{F}_{\mathcal{R}}^1$ has at least one center $u \in U_0$. So, consider the following procedure to construct all star-factors of $\mathcal{F}_{\mathcal{R}}^1$:

1. Choose a center $u \in U_0$; there are at most $n$ possibilities to do this.
2. For the center $u$, do the following:

    (a) Choose a set of $k \leqslant m$ vertices in $V_u$ and connect them to $u$.
    (b) Choose a set of $d - k$ vertices in $[n] \setminus (V_u \cup \{u\})$ and connect them to $u$.

3. Choose a subset of $m - 1$ distinct centers from the remaining $n - (d+1)$ vertices. There are at most $\binom{n-d-1}{m-1} \leqslant \binom{n-1}{m-1} = \frac{m}{n}\binom{n}{m}$ possibilities to do this.
4. Choose a partition of the remaining $n - m - d$ vertices into $m - 1$ blocks of size $d$, and connect the $i$th largest of the $m - 1$ chosen centers to all vertices in the $i$th block. There are at most $\binom{n-m-d}{d,\dots,d} = \frac{(n-m-d)!}{d!^{m-1}}$ possibilities to do this.                                                                      □

**Claim 3.21** *Every star-factor $F \in \mathcal{F}_{\mathcal{R}}^1$ can be produced by the above procedure.*

**Proof** Take a star-factor $F \in \mathcal{F}_{\mathcal{R}}^1$ containing a star $S_u \subseteq F$ centered in a vertex $u \in U_0$. The center $u$ can be picked by Step 1 of the procedure. By Claim 3.20, the star $S_u$ can only have $k \leqslant m$ leaves in $V_u$, and Step 2(a) of the procedure can pick all these $k$ leaves of $S_u$. The remaining $d - k$ leaves of the star $S_u$ must belong to the set $[n] \setminus (V_u \cup \{u\})$. So, Step 2(b) can pick these $d - k$ leaves of $S_u$. Since the remaining two steps 3 and 4 of the procedure can construct *any* star-factor of $K_n \setminus S_u$, the rest of the star-factor $F$ can be constructed by these steps.                                      □

To upper-bound the number of possibilities in Step 2, we will use Lemma 3.18. We have a fixed subset $V_u$ of $[n]$ size $|V_u| \geqslant pn$ for $p := 1/16$. The set $S$ of leaves of each star $S_u$ centered in $u$ is a $d$-element subset of $[n]$ such that $|S \cap V_u| \leqslant m \leqslant d/32 = (p - \epsilon)d$ for $\epsilon := 1/32$. Thus, by Lemma 3.18, the number of possibilities in Step 2 is at most

$$\gamma \cdot \binom{n}{d} \quad \text{where} \quad \gamma := e^{-2\epsilon^2 d} \leqslant \boxed{2^{-\Omega(d)}} .$$

From the first inequality of Eq. (3.4) (applied with $x := m$, $a := n$, and $b := d$), we have

$$\binom{n}{d} \leqslant \left(\frac{n-m}{n-d-m}\right)^m \binom{n-m}{d} \leqslant C \cdot \binom{n-m}{d} ,$$

where $C = \left(1 + \frac{d}{n-d-m}\right)^m \leqslant \exp\left(\frac{md}{n-d-m}\right)$ is a constant since $md = \mathcal{O}(n)$ and $d + m = o(n)$. Thus, the total number of possibilities in all steps 1–4 and, by Claim 3.21, also the number $|\mathcal{F}_{\mathcal{R}}^1|$ of star-factors in $\mathcal{F}_{\mathcal{R}}^1$ are at most a constant times

$$\underbrace{\gamma \cdot n \binom{n-m}{d}}_{\text{Steps 1 and 2}} \underbrace{\frac{m}{n}\binom{n}{m}}_{\text{Step 3}} \underbrace{\frac{(n-m-d)!}{d!^{m-1}}}_{\text{Step 4}} = \gamma \cdot m \underbrace{\binom{n}{m}\frac{(n-m)!}{d!^m}}_{= |\mathcal{F}|} = \gamma \cdot m|\mathcal{F}|.$$

Recall that we have only used the conditions $(d + 1)m = n$, $m \leqslant d/32$, and $m = \Theta(\sqrt{n})$ on the parameters $m$ and $d$. By taking $d := 6\sqrt{n}$ and $m := n/(d + 1)$, all three conditions on the parameters $m$ and $d$ are fulfilled, and the desired upper bound $|\mathcal{F}_{\mathcal{R}}^1| \leqslant m|\mathcal{F}| \cdot 2^{-\Omega(d)}$ follows.

*Proof of Claim 3.20* Fix a balanced rectangle $\mathcal{R} = \mathcal{A} \vee \mathcal{B}$ lying in the family $\mathcal{T}_n$ of all spanning trees of $K_n$. Fix a spanning tree $T_0 = A_0 \cup B_0$ with $A_0 \in \mathcal{A}$ and $B_0 \in \mathcal{B}$ containing some star-factor $F_0$. Hence, $T_0$ consists of $m$ vertex-disjoint $d$-stars of $F_0$ and additional $m - 1$ edges joining these stars into the spanning tree $T_0$. The spanning tree $T_0 = A_0 \cup B_0$ defines a complete bipartite subgraph $U \times V$ of $K_n$ as follows. Let $c_1, \ldots, c_m$ be the centers of the stars of $F_0$. Every remaining vertex $v$ of $K_n$ is connected in the tree $T_0$ by a *unique* edge $e_v$ to one of the centers $c_1, \ldots, c_m$. This gives us a partition $U \cup V$ of the vertices in $[n] \setminus \{c_1, \ldots, c_m\}$ into two disjoint sets $U = \{v : e_v \in A_0\}$ and $V = \{v : e_v \in B_0\}$ determined by the forests $A_0$ and $B_0$. That is, we partite the leaves $v$ of the $m$ stars of the star-factor $F_0$ into two parts depending on whether the unique edge of the corresponding star belongs to the forest $A_0$ or to the forest $B_0$. By the balancedness of the rectangle $\mathcal{R} = \mathcal{A} \vee \mathcal{B}$, we have $|A_0|, |B_0| \geqslant (n - 1)/3$; hence, both $|U|$ and $|V|$ are at least $(n - 1)/3 - m \geqslant n/4$ since $m = o(n)$.

Call an edge $e$ of $U \times V$ an $\mathcal{A}$-*edge* (respectively, a $\mathcal{B}$-*edge*) if $e$ belongs to some forest $A \in \mathcal{A}$ (respectively, to some forest $B \in \mathcal{B}$). No $\mathcal{A}$-edge can be a $\mathcal{B}$-edge and vice versa: forests $A \in \mathcal{A}$ and $B \in \mathcal{B}$ are edge-disjoint. We can assume that at least half of the edges in $U \times V$ are $\mathcal{A}$-edges; otherwise consider $\mathcal{B}$-edges.

For a vertex $u \in U$, let $V_u \subseteq V$ denote the set of all vertices $v \in V$ joined to $u$ by $\mathcal{A}$-edges. Consider the set of vertices $U_0 := \{u \in U : |V_u| \geqslant \frac{1}{4}|V|\}$. Since $|V| \geqslant n/4$, every vertex $u \in U_0$ has $|V_u| \geqslant n/16$ incident $\mathcal{A}$-edges. Our goal is to show that the set $U_0$ together with the corresponding sets $V_u \subseteq V$ associated with the vertices $u \in U_0$ has the properties stated in Claim 3.20. This is done by the following two simple claims. $\square$

**Claim 3.22** $|U_0| \geqslant n/16$.

**Proof** Since $|U| \geqslant n/4$, it is enough to show that $|U_0| \geqslant \frac{1}{4}|U|$. We know that at least a half of the edges of $U \times V$ are $\mathcal{A}$-edges. By the definition, every vertex $u \in U \setminus U_0$ can be incident to $|V_u| < \frac{1}{4}|V|$ $\mathcal{A}$-edges of $U \times V$. Since clearly $|V_u| \leqslant |V|$ holds for all vertices $u \in U_0$, $|U_0| < \frac{1}{4}|U|$ would imply that *fewer* than $|U| \cdot \frac{1}{4}|V| + \frac{1}{4}|U| \cdot |V| = \frac{1}{2}|U \times V|$ edges of $U \times V$ are $\mathcal{A}$-edges, a contradiction. $\square$

**Claim 3.23** *Let $S_u$ be a $d$-star covered by the rectangle $\mathcal{R}$. If the center $u$ of $S_u$ belongs to $U$, then $|S_u \cap (\{u\} \times V_u)| \leqslant m$.*

**Proof** Since the star $S_u$ is covered by the rectangle $\mathcal{R}$, $S_u \subseteq A \cup B$ holds for some forests $A \in \mathcal{A}$ and $B \in \mathcal{B}$. Since all edges in $\{u\} \times V_u$ are $\mathcal{A}$-edges, while those in $B$ are $\mathcal{B}$-edges, we have $B \cap (\{u\} \times V_u) = \emptyset$. So, all edges of $S_u \cap (\{u\} \times V_u)$ belong to the forest $A$, and it remains to show the (even stronger) upper bound $|A \cap (\{u\} \times V_u)| \leqslant m$.

So, take a vertex $u \in U$, and assume for a contradiction that there are $l \geqslant m + 1$ vertices $v_1, \ldots, v_l$ in $V$ such that all edges $\{u, v_1\}, \ldots, \{u, v_l\}$ belong to the forest $A$. In the (fixed) spanning tree $T_0 = A_0 \cup B_0$ (determining the partition $U \cup V$ of vertices in $[n] \setminus \{c_1, \ldots, c_m\}$), each of these $l$ vertices is joined by an edge of the forest $B_0$ to one of the $m$ centers $c_1, \ldots, c_m$ of stars of the star-factor contained in $T_0$. Since $l > m$, some two of these vertices $v_i$ and $v_j$ must be joined in $B_0$ to the *same* center $c \in \{c_1, \ldots, c_m\}$ (by the pigeonhole principle). Since all graphs in the rectangle must be spanning trees, the graph $A \cup B_0$ must also be a (spanning) tree. But the edges $\{u, v_i\}$ and $\{u, v_j\}$ of $A$ together with edges $\{v_i, c\}, \{v_j, c\}$ of $B_0$ form a cycle $u - v_i - c - v_j - u$ in $A \cup B_0$, a contradiction. □

## 3.6   Coefficients Can Matter

As mentioned in Sect. 1.12, almost all known lower bounds on the monotone arithmetic $(+, \times)$ circuit complexity of polynomials $f(x) = \sum_{a \in A} c_a \prod_{i=1}^{n} x_i^{a_i}$ are actually lower bounds on the Minkowski circuit complexity $\mathrm{L}(A)$ of their sets $A \subseteq \mathbb{N}^n$ of exponent vectors. That is, these arguments exploit the structure of monomials and completely ignore their coefficients. Yehudayoff [19] has shown that strong lower bounds can be proved even for polynomials whose sets of exponent vectors have very simple structure. His polynomial $f$ has $n^2$ variables $x_{i,j}$ arranged into an $n \times n$ matrix and is of the form

$$f(x) = \sum_{\varphi} 2^{-|\varphi([n])|} \prod_{i=1}^{n} x_{i, \varphi(i)} \,, \tag{3.5}$$

where the sum is taken over all $n^n$ mappings $\varphi : [n] \to [n]$.

We recall Theorem 1.38:

**Theorem 1.38** (Yehudayoff [19]) $\mathtt{Arith}(f) = 2^{\Omega(n/\log n)}$.

To prove the theorem, let us first fix some notation. Until the rest of this section, let $f$ stand for Yehudayoff's polynomial (3.5).

We will only consider polynomials over variables $x_{i,j}$ with $i, j \in [n]$. We focus on monomials of the form $\alpha = \prod_{i \in I} x_{i, \varphi(i)}$, where $I \subseteq [n]$ and $\varphi : [n] \to [n]$. We call $I(\alpha) := I$ the *domain* and $J(\alpha) := \varphi(I) = \{\varphi(i) : i \in I\}$ the *range* of the monomial $\alpha$. For a polynomial $g$, $I(g)$ denotes the union of the domains $I(\alpha)$

of all monomials $\alpha \in \mathrm{mon}(g)$ of $g$. Say that a polynomial $g$ is *ordered* if all its monomials $\alpha$ have the same domain $I(\alpha) = I(g)$. Note that our polynomial $f$ is ordered with $I(f) = [n]$. By some abuse of notation, for a monomial $\alpha$, we will denote by $g(\alpha)$ the *coefficient* of the monomial $\alpha$ in the polynomial $g$. That is, if a monomial $\alpha = \prod_{i \in I} x_{i,\varphi(i)}$ appears in the polynomial $g$ as a term $c \prod_{i \in I} x_{i,\varphi(i)}$, then $g(\alpha) = c$. For polynomials $g$ and $h$, we write $g \leqslant h$ if $g(\alpha) \leqslant h(\alpha)$ holds for all monomials $\alpha$.

Theorem 1.38 is the direct consequence of the following two lemmas.

**Lemma 3.24** *If* $s = \mathtt{Arith}(f)$ *and* $n > 3$, *then the polynomial* $f$ *can be written as the sum* $f = g_1 h_1 + \cdots + g_s h_s$ *of products of ordered polynomials with* $0 \leqslant g_i h_i \leqslant f$, $n/3 \leqslant |I(g_i)| \leqslant 2n/3$ *and* $I(h_i) = [n] \setminus I(g_i)$.

*Proof* We are going to apply the decomposition lemma for polynomials (Lemma 3.1). Recall that a *component* of a polynomial $f$ is a polynomial $g$ such that $f = gh + q$ holds for some polynomials $h$ and $q$. In particular, components of Yehudayoff's polynomial $f$ are polynomials $g$ and $h$ whose monomials are of the form $\alpha = \prod_{i \in I(\alpha)} x_{i,\varphi(i)}$, where $I(\alpha) \subseteq [n]$ and $\varphi : [n] \to [n]$. Since the polynomial $f$ is ordered, the polynomials $gh \leqslant f$ are also ordered, i.e., $I(\alpha) = I(g)$ and $I(\beta) = I(h)$ holds for all monomials $\alpha$ of $g$ and all monomials $\beta$ of $h$. So, consider the measure $\mu(g) := |I(g)|$ of components $g$ of $f$. Since $\mu(x_{i,j}) = 1$ for each of the variables $x_{i,j}$, the measure $\mu$ is normalized. We also have $I(g + h) = I(g) = I(h)$, and $I(gh) = |I(g)| + |I(h)|$ with $I(g) \cap I(h) = \emptyset$ in the latter case. This, in particular, implies that the measure is subadditive, that is, both $\mu(g + h)$ and $\mu(gh)$ are at most $\mu(g) + \mu(h)$.

By Lemma 3.1 (applied with this measure $\mu$), the polynomial $f$ can be written as the sum $f = g_1 h_1 + \cdots + g_s h_s$ of products of polynomials with $0 \leqslant g_i h_i \leqslant f$, $n/3 \leqslant |I(g_i)| \leqslant 2n/3$. Since $g_i \cdot h_i \leqslant f$ and the polynomial $f$ is ordered and multilinear, the polynomials $g_i$ and $h_i$ are ordered, and $I(h_i) = I(f) \setminus I(g_i) = [n] \setminus I(g_i)$. $\square$

Let $n$ be sufficiently large, and let $\delta := \Theta(1/\log n)$ be such that $\delta \leqslant 1/20$ and the parameter

$$m := \delta n = \Theta(n/\log n)$$

is an integer. Define the *weight* $\|p\|$ of a polynomial $0 \leqslant p \leqslant f$ as the sum

$$\|p\| := \sum_{\substack{\alpha \in \mathrm{mon}(p) \\ |J(\alpha)| = m}} p(\alpha)$$

of all coefficients $p(\alpha)$ to those monomials $\alpha$ of $p$ whose range $J(\alpha)$ has exactly $m$ elements. Note that $\|p + q\| \leqslant \|p\| + \|q\|$.

**Lemma 3.25** *Let* $g$ *and* $h$ *be ordered polynomials such that* $0 \leqslant gh \leqslant f$, $n/3 \leqslant |I(g)| \leqslant 2n/3$, *and* $I(h) = [n] \setminus I(g)$. *Then* $\|gh\| \leqslant 2^{-\Omega(m)} \cdot \|f\|$.

Note that Lemmas 3.24 and 3.25 already yield Theorem 1.38. Let $s = \text{Arith}(f)$. Then, by Lemma 3.24, the polynomial $f$ can be written as the sum $f = \sum_{i=1}^{s} g_i h_i$ of products of ordered polynomials with $0 \leqslant g_i h_i \leqslant f$, $n/3 \leqslant |I(g_i)| \leqslant 2n/3$, and $I(h_i) = [n] \setminus I(g_i)$. So, by Lemma 3.25, we have $\|f\| \leqslant \sum_{i=1}^{s} \|g_i h_i\| \leqslant s \cdot 2^{-\Omega(m)} \cdot \|f\|$, from which the desired lower bound $s \geqslant 2^{\Omega(m)} = 2^{\Omega(n/\log n)}$ follows.

*Proof of Lemma 3.25* First, note that the weight of the entire Yehydayoff's polynomial $f$ is

$$\|f\| = \sum_{\substack{|S|=m}} \sum_{\substack{\varphi:[n]\to[n] \\ \varphi([n])=S}} 2^{-|\varphi([n])|} = \binom{n}{m} 2^{-m} F_{n,m}, \qquad (3.6)$$

where $F_{n,m}$ is the number of *onto* mappings from $[n]$ to $[m]$. This latter number can be estimated as follows:

$$\tfrac{1}{2} m^n \leqslant F_{n,m} \leqslant m^n. \qquad (3.7)$$

The right inequality is obvious. To show the left inequality, observe that $m^n - F_{n,m}$ is at most $m(m-1)^n$: there are $m$ possibilities to avoid a single value in $[m]$. Using $1 - t \leqslant e^{-t}$, we obtain $m(m-1)^n = m(1 - 1/m)^n m^n \leqslant m e^{-n/m} m^n$. This is at most $m^n/2$ if $2m \leqslant e^{n/m}$, which holds for our particular choice of $m$ because $n/m \geqslant 1 + \log n$. $\qquad \square$

Since $0 \leqslant gh \leqslant f$, the coefficients $g(\alpha)h(\beta)$ to monomials $\alpha\beta$ of $gh$ satisfy

$$0 \leqslant g(\alpha)h(\beta) \leqslant f(\alpha\beta) = 2^{-|J(\alpha\beta)|} = 2^{-|J(\alpha)\cup J(\beta)|}. \qquad (3.8)$$

We are interested in the sum

$$\|gh\| = \sum_{\substack{|S|=m}} \sum_{\substack{J(\alpha\beta)=S}} g(\alpha)h(\beta), \qquad (3.9)$$

where the second sum is over all monomials $\alpha$ of $g$ and all monomials $\beta$ of $h$ such that $J(\alpha\beta) = S$. Our goal is to show that $\|gh\| \leqslant 2^{-\Omega(m)} \cdot \|f\|$. For this, by Eq. (3.6), it is enough to show that $\|gh\| \leqslant 2^{-\Omega(m)} \cdot \binom{n}{m} 2^{-m} F_{n,m}$ holds.

Let us note that in Eq. (3.9), we are only interested in coefficients at monomials $\alpha\beta$ of the polynomial $gh$ whose range $J(\alpha\beta) = J(\alpha) \cup J(\beta)$ is of size $|J(\alpha\beta)| = m$. In particular, we have only to consider monomials $\alpha$ and $\beta$ whose ranges $J(\alpha)$ and $J(\beta)$ have at most $m$ elements. Depending on how near to the maximal possible value $m$ the sizes $|J(\alpha)|$ of their ranges are, we divide the monomials $\alpha$ into "narrow" and "wide." Namely, we take the parameter

$$r := (1 - \varepsilon)m \quad \text{for} \quad \varepsilon := 1/20$$

and call a monomial $\alpha$ is *narrow* if $0 \leqslant |J(\alpha)| < r$ and *wide* if $r \leqslant |J(\alpha)| \leqslant m$.

We split the sum (3.9) into two sums, one over narrow monomials and the other over wide monomials. Namely, for a set $S \subseteq [n]$, let Narrow($S$) be the sum of coefficients $g(\alpha)h(\beta)$ over all pairs $(\alpha, \beta)$ of monomials $\alpha \in \text{mon}(g)$ and $\beta \in \text{mon}(h)$ such that $J(\alpha\beta) = S$ and $|J(\alpha)| < r$ or $|J(\beta)| < r$ (at least one of the two monomials is narrow). Similarly, let Wide($S$) be the sum of coefficients $g(\alpha)h(\beta)$ over all pairs $(\alpha, \beta)$ of monomials such that $J(\alpha\beta) = S$ and $r \leqslant |J(\alpha)|, |J(\beta)| \leqslant m$ (both monomials are wide). Note that

$$\|gh\| = \text{Narrow}(m) + \text{Wide}(m), \tag{3.10}$$

where $\text{Narrow}(m) := \sum_{|S|=m} \text{Narrow}(S)$ and $\text{Wide}(m) := \sum_{|S|=m} \text{Wide}(S)$.

Let us first show that the contribution to $\|gh\|$ of narrow monomials is small. This comes just from the fact that the total number of such monomials is small.

**Claim 3.26** $\text{Narrow}(m) \leqslant 2^{-\Omega(n)} \cdot \binom{n}{m} 2^{-m} F_{n,m} = 2^{-\Omega(n)} \cdot \|f\|$.

**Proof** By symmetry, it is enough to show that the sum of $g(\alpha)h(\beta)$ over all narrow monomials $\alpha \in \text{mon}(g)$ and *all* monomials $\beta \in \text{mon}(h)$ with $J(\alpha\beta) = S$ is at most $2^{-\Omega(n)} \cdot 2^{-m} F_{n,m}$; then Narrow($S$) is at most twice this number. The number of possibilities to choose the range $J(\alpha) \subseteq S$ for a narrow monomial $\alpha$ is $\binom{m}{<r} = \sum_{i=0}^{r-1} \binom{m}{i} \leqslant 2\binom{m}{r} \leqslant 2(em/r)^r = 2[e/(1-\varepsilon)]^{(1-\varepsilon)\delta n}$. Thus, since $\delta \leqslant 1/\log n \to 0$ while $\varepsilon = 1/20$ is a constant, we have $\binom{m}{<r} \ll e^{-\varepsilon n/3}$.

After a set $J(\alpha)$ of size $|J(\alpha)| < r$ is chosen, there are at most $r^{|I(g)|}$ possibilities to choose a mapping $\varphi : I(g) \to J(\alpha)$ and at most $m^{|I(h)|}$ possibilities to choose a mapping $\psi : I(h) \to S$. We know that $|I(g)|, |I(h)| \geqslant n/3$ and $|I(g)|+|I(h)| = n$. Thus, since $r = (1 - \varepsilon)m$ and $1 - \varepsilon \leqslant e^{-\varepsilon}$, we have

$$\sum_{\substack{J(\alpha\beta)=S \\ |J(\alpha)|<r}} g(\alpha)h(\beta) \overset{(3.8)}{\leqslant} \sum_{\substack{J(\alpha\beta)=S \\ |J(\alpha)|<r}} 2^{-m} \leqslant \binom{m}{<r} r^{|I(g)|} m^{|I(h)|} 2^{-m}$$

$$= \binom{m}{<r} (1-\varepsilon)^{n/3} m^n 2^{-m} \leqslant \binom{m}{<r} e^{-\varepsilon n/3} \cdot 2 F_{n,m} \cdot 2^{-m},$$

which is at most $2^{-\Omega(n)} \cdot 2^{-m} F_{n,m}$ because $\binom{m}{<r} \ll e^{-\varepsilon n/3}$. $\qquad\square$

To complete the proof of Lemma 3.25, it remains to upper-bound the second term Wide($m$) in Eq. (3.10). That is, it remains to estimate the contribution to the sum (3.9) of pairs $(\alpha, \beta)$, where *both* monomials $\alpha$ and $\beta$ are wide. To do this, we will consider two cases depending on whether some of these wide monomials have "large" coefficients or not. For this, let $\|g\|_\infty$ be the maximum coefficient $g(\alpha)$ of a wide monomial $\alpha$ of $g$ and similarly for the polynomial $h$. By scaling the polynomials $g$ and $h$ by scalars, we can assume that

$$\|g\|_\infty = \|h\|_\infty. \tag{3.11}$$

Indeed, if $\|g\|_\infty = 0$ or $\|h\|_\infty = 0$, then either $g$ or $h$ has no wide monomials, and by Claim 3.26, Lemma 3.25 holds in this case. So, we can assume that $\|g\|_\infty > 0$ and $\|h\|_\infty > 0$. Then, taking the constant $c := \sqrt{\frac{\|h\|_\infty}{\|g\|_\infty}} > 0$, we get $\|c \cdot g\|_\infty = \sqrt{\|g\|_\infty \cdot \|h\|_\infty} = \left\|\frac{1}{c} \cdot h\right\|_\infty$. So, if $\|g\|_\infty \neq \|h\|_\infty$, then, instead of the product $gh$, we can consider the product $g'h'$ of polynomials $g' := c \cdot g$ and $h' := \frac{1}{c} \cdot h$; then $\|g'\|_\infty = \|h'\|_\infty$ and $\|gh\| = \|g'h'\|$ hold. Note that for a monomial to be wide or narrow, its coefficient is irrelevant.

Recall that our two parameters are $m := \delta n$ and $r := m - \varepsilon m$ with $\delta = \Theta(1/\log n)$ and $\varepsilon = 1/20$.

*Case 1* $\|g\|_\infty \leqslant 2^{-r}$. By Eq. (3.11), we then also have $\|h\|_\infty \leqslant 2^{-r}$. Hence, in this case, we have $g(\alpha)h(\beta) \leqslant 2^{-2r}$ for all pairs $(\alpha, \beta)$ of wide monomials. Thus, for any set $S \subseteq [n]$ of size $|S| = m$, Wide$(S)$ is at most $2^{-2r}$ times the total number $m^n$ of all mappings $\varphi : [n] \to S$, i.e., is at most

$$\boxed{2^{-2r}\, m^n} \leqslant 2^{-2r} \cdot 2 \cdot F_{n,m} = 2^{-m+2\varepsilon m+1} \cdot 2^{-m} F_{n,m} \leqslant \boxed{2^{-\Omega(m)}} \cdot 2^{-m} F_{n,m}\,,$$

where the first inequality follows from Eq. (3.7). Finally, by summing over all $\binom{n}{m}$ subsets $S \subset [n]$ of size $|S| = m$, the desired upper bound Wide$(m) \leqslant 2^{-\Omega(m)} \cdot \|f\|$ follows. This completes the proof of Lemma 3.25 in Case 1.

*Case 2* $\|g\|_\infty \geqslant 2^{-r}$. By Eq. (3.11), we then also have $\|h\|_\infty \geqslant 2^{-r}$. Thus, in this case, there is a monomial $\alpha_0$ of $g$ and a monomial $\beta_0$ of $h$ such that $r \leqslant |J(\alpha_0)|, |J(\beta_0)| \leqslant m$ and $g(\alpha_0), h(\beta_0) \geqslant 2^{-r}$. The sum Wide$(m) = \sum_S$ Wide$(S)$ is over all sets $S \in \binom{[n]}{m}$. Using these two monomials $\alpha_0$ and $\beta_0$, we split the sum Wide$(m)$ into two parts, where the first part consists of the sets $S$ in the family

$$\mathcal{F} := \left\{S \subseteq [n] : |S| = m \text{ and } \big|S \setminus J(\alpha_0\beta_0)\big| < r - 2\delta m\right\}$$

of potential $m$-element ranges $J(\alpha\beta) = S$ of monomials of $gh$ sharing "many" elements with the range $J(\alpha_0\beta_0) \subseteq [n]$ of the monomial $\alpha_0\beta_0$ of $gh$. Let us first show that $|\mathcal{F}|$ is a relatively small fraction of all sets $S \in \binom{[n]}{m}$.

**Claim 3.27** $|\mathcal{F}| \leqslant \boxed{2^{-\Omega(m)}} \binom{n}{m}$.

*Proof* This upper bound is proved in [19] using the Chernoff–Hoeffding inequality. It is also a direct consequence of Lemma 3.18 stating that if $R \subseteq [n]$ is a fixed set with $|R| \geqslant pn$ elements, then for every $0 \leqslant \epsilon \leqslant p$, the family of $m$-element subsets $S$ of $[n]$ satisfying $|S \cap R| \leqslant (p - \epsilon)m$ can have at most $e^{-2\epsilon^2 m}\binom{n}{m}$ sets.

In our case, we have a fixed subset $R := [n] \setminus J(\alpha_0\beta_0)$ of $[n]$ of size $|R| \geqslant n - 2m = (1 - 2\delta)n = pn$ with $p := 1 - 2\delta$. The family $\mathcal{F}$ consists of all $m$-element subsets $S \subseteq [n]$ such that $|S \cap R| = |S \setminus J(\alpha_0\beta_0)| < r - 2\delta m = (1 - \varepsilon - 2\delta)m = (p - \varepsilon)m$. So, Lemma 3.18 yields $|\mathcal{F}| \leqslant e^{-2\varepsilon^2 m}\binom{n}{m} \leqslant e^{-\Omega(m)}\binom{n}{m}$. $\qquad\square$

Together with the trivial upper bound $g(\alpha)h(\beta) \leqslant 2^{-m}$ given by Eq. (3.8), Claim 3.27 yields

$$\sum_{S \in \mathcal{F}} \text{Wide}(S) \leqslant \sum_{S \in \mathcal{F}} \sum_{J(\alpha\beta)=S} 2^{-m} \leqslant |\mathcal{F}| 2^{-m} m^n \stackrel{\text{Clm. 3.27}}{\leqslant} 2^{-\Omega(m)} \cdot \binom{n}{m} 2^{-m} m^n ,$$

which, by Eq. (3.6), is at most $2^{-\Omega(m)} \| f \|$. It thus remains to show that also the sum $\sum_{S \notin \mathcal{F}} \text{Wide}(S)$ over all sets $S \in \binom{[n]}{m} \setminus \mathcal{F}$ does not exceed $2^{-\Omega(m)} \| f \|$. So, fix such a set $S \notin \mathcal{F}$. Recall that $\text{Wide}(S)$ is the sum of $g(\alpha)h(\beta)$ over all pairs $(\alpha, \beta)$ of wide monomials satisfying $J(\alpha) \cup J(\beta) = S$.

Together with a trivial upper bound $g(\alpha_0)h(\beta) \leqslant 2^{-|J(\alpha_0\beta)|}$ given by Eq. (3.8), our assumption $2^{-r} \leqslant g(\alpha_0)$ yields $h(\beta) \leqslant 2^{-|J(\alpha_0\beta)|+r}$ for all monomials $\beta \in$ mon$(h)$, including all wide monomials $\beta$ with $J(\beta) \subseteq S$. For every such monomial $\beta$, the set $S \setminus J(\alpha_0)$ is contained in the union of sets $S \setminus J(\beta)$ and $J(\beta) \setminus J(\alpha_0)$:

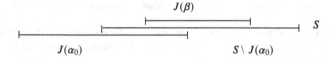

This yields $|J(\beta) \setminus J(\alpha_0)| \geqslant |S \setminus J(\alpha_0)| - |S \setminus J(\beta)|$, where $|S \setminus J(\beta)| \leqslant m - r$ because $J(\beta) \subseteq S$ and $|J(\beta)| \geqslant r$ (monomial $\beta$ is wide), and $|S \setminus J(\alpha_0)| \geqslant |S \setminus J(\alpha_0\beta_0)| \geqslant r - 2\delta m$ because $S \notin \mathcal{F}$. Since the monomial $\alpha_0$ is wide, we also have $|J(\alpha_0)| \geqslant r$. Thus, $|J(\alpha_0\beta)| = |J(\alpha_0)| + |J(\beta) \setminus J(\alpha_0)| \geqslant r + (r - 2\delta m) - (m - r) = 3r - 2\delta m - m$. Since $r = m - \varepsilon m$ and $\varepsilon + \delta \leqslant 1/10$, we obtain

$$h(\beta) \leqslant 2^{-|J(\alpha_0\beta)|+r} \leqslant 2^{-2r+2\delta m+m} = 2^{-m+2(\varepsilon+\delta)m} \leqslant 2^{-4m/5} .$$

By considering the monomial $\beta_0$ instead of $\alpha_0$, the same upper bound $g(\alpha) \leqslant 2^{-4m/5}$ holds for each monomial $\alpha$ of $g$ with $J(\alpha) \subseteq S$ and $|J(\alpha)| \geqslant r$. Hence, $g(\alpha)h(\beta) \leqslant 2^{-8m/5}$ holds for every pair $(\alpha, \beta)$ of wide monomials with $J(\alpha\beta) = S$. The sum $\text{Wide}(S)$ of $g(\alpha)h(\beta)$ over all such pairs $(\alpha, \beta)$ is thus at most $2^{-8m/5}$ times the total number $F_{n,m}$ of onto mappings $\varphi : [n] \to S$. Hence, for every set $S \in \binom{[n]}{m} \setminus \mathcal{F}$, we have

$$\text{Wide}(S) \leqslant 2^{-8m/5} F_{n,m} = 2^{-3m/5} \cdot 2^{-m} F_{n,m} .$$

Finally, we sum over all (at most $\binom{n}{m}$) sets $S \notin \mathcal{F}$:

$$\sum_{S \notin \mathcal{F}} \text{Wide}(S) \leqslant 2^{-3m/5} \cdot \binom{n}{m} 2^{-m} F_{n,m} \stackrel{(3.6)}{\leqslant} 2^{-\Omega(m)} \| f \| .$$

This completes the proof of Lemma 3.25 and, hence, also of Theorem 1.38.   $\square$

# References

1. Alon, N., Spencer, J.H.: The Probabilistic Method, 2nd edn. John Wiley, Hoboken (2000)
2. Bellman, R.: Dynamic programming treatment of the Travelling Salesman problem. J. ACM **9**(1), 61–63 (1962)
3. Bollobás, B.: Random Graphs, 2nd edn. Cambridge University Press, Cambridge (2001)
4. Chvátal, V.: The tail of the hypergeometric distribution. Discrete Math. **25**, 285–287 (1979)
5. Edmonds, J.: Optimum branchings. J. Res. Nat. Bur. Stand. **71B**(4), 233–240 (1967)
6. Erdős, P., Spencer, J.: Lopsided Lovász local lemma and Latin transversals. Discrete Appl. Math. **30**(2–3), 151–154 (1991)
7. Held, M., Karp, R.M.: A dynamic programming approach to sequencing problems. SIAM J. Appl. Math. **10**, 196–210 (1962)
8. Hoeffding, W.: Probability inequalities for sums of bounded random variables. J. Am. Stat. Assoc. **58**(301), 13–30 (1963)
9. Hyafil, L.: On the parallel evaluation of multivariate polynomials. SIAM J. Comput. **8**(2), 120–123 (1979)
10. Jerrum, M., Snir, M.: Some exact complexity results for straight-line computations over semirings. J. ACM **29**(3), 874–897 (1982)
11. Jukna, S., Seiwert, H.: Greedy can beat pure dynamic programming. Inf. Process. Lett. **142**, 90–95 (2019)
12. Korhonen, T.: Lower bounds on dynamic programming for minimum weight independent set. In: Proceedings of 48th International Colloquium on Automata, Languages and Programming (ICALP) (2021)
13. Kruskal, J.B.: On the shortest spanning subtree of a graph and the traveling salesman problem. Proc. AMS **7**, 48–50 (1956)
14. Moon, J.W.: Counting labelled trees. In: Canadian Mathematical Monographs, vol. 1 (1970)
15. Raz, R., Yehudayoff, A.: Multilinear formulas, maximal-partition discrepancy and mixed-sources extractors. J. Comput. Syst. Sci. **77**(1), 167–190 (2011)
16. Robertson, N., Seymour, P.D.: Graph minors II: algorithm aspects of tree-width. J. Algorithms **7**(3), 309–322 (1986)
17. Seymour, P.D., Thomas, R.: Graph searching and a min-max theorem for tree-width. J. Combin. Theory Ser. B **58**, 22–33 (1993)
18. Valiant, L.G.: Negation can be exponentially powerful. Theor. Comput. Sci. **12**, 303–314 (1980)
19. Yehudayoff, A.: Separating monotone VP and VNP. In: Proceedings of 51st Annual ACM SIGACT Symposium on Theory of Computing, STOC, pp. 425–429. ACM, New York (2019)

# Chapter 4
# Bounds for Approximating Circuits

**Abstract** We have seen that some optimization problems cannot be solved by tropical circuits using a polynomial number of gates. But what happens if we allow the circuits to only output "almost optimal" values: can at least then the circuit size be substantially reduced? The goal of this chapter is to present recently emerged arguments for proving lower bounds on the size of approximating tropical circuits. It turns out that the task of proving lower bounds for approximating $(\min, +)$ circuits is much easier than for $(\max, +)$ circuits. While the size of the former circuits is lower bounded by the size of monotone Boolean circuits, approximating $(\max, +)$ circuits are much more powerful. For them, neither the Boolean bound nor even counting arguments work. Still, also for $(\max, +)$ circuits, the situation is not hopeless: there is a general combinatorial bound leading to strong lower bounds on the size of such circuits approximating explicit maximization problems, as well.

## 4.1 Approximating (Min, +) Circuits

Recall that a tropical circuit $\Phi$ *approximates* a given optimization problem $f$ within a *factor* $r \geqslant 1$ if for every input weighting $x \in \mathbb{R}_+^n$, the output value $\Phi(x)$ of the circuit lies:

o Between $f(x)$ and $r \cdot f(x)$, in the case when $f$ is a *minimization* problem.
o Between $f(x)/r$ and $f(x)$, in the case when $f$ is a *maximization* problem.

For a finite set $A \subseteq \mathbb{N}^n$ of vectors and a real number $r \geqslant 1$, $\text{Min}_r(A)$ denotes the minimum size of a tropical $(\min, +)$ circuit approximating the *minimization* problem $f(x) = \min_{a \in A} \langle a, x \rangle$ on $A$ within the factor $r$. The corresponding complexity measure in the case of *maximization* problems and $(\max, +)$ circuits is denoted by $\text{Max}_r(A)$.

Recall that, in the case of combinatorial optimization (or $0/1$ optimization), the set $A \subseteq \{0, 1\}^n$ of feasible solutions consists of 0-1 vectors. Thus, every such problem is either the function $f(x) = \min_{S \in \mathcal{F}} \sum_{i \in S} x_i$ or the function $f(x) = \min_{S \in \mathcal{F}} \sum_{i \in S} x_i$, where $\mathcal{F} = \{\sup(a) \colon a \in A\}$. The problem itself is to compute or approximate the values $f(x)$ for all input weightings $x \in \mathbb{R}_+^n$. In the

© The Author(s), under exclusive license to Springer Nature Switzerland AG 2023
S. Jukna, *Tropical Circuit Complexity*, SpringerBriefs in Mathematics,
https://doi.org/10.1007/978-3-031-42354-3_4

case, when feasible solutions are given as *sets*, we will write $\text{Min}_r(\mathcal{F})$ and $\text{Max}_r(\mathcal{F})$ instead of $\text{Min}_r(A)$ and $\text{Max}_r(A)$, where $A$ is the set of characteristic 0-1 vectors of sets $S \in \mathcal{F}$. As before, we will freely switch between the vector and set-theoretic languages, whichever of them is more convenient in a particular context. Note that the larger the allowed approximation factor is, the smaller circuits are required to approximate a given optimization problem: if $s \geqslant r$, then $\text{Min}_s(\mathcal{F}) \leqslant \text{Min}_r(\mathcal{F})$ and $\text{Max}_s(\mathcal{F}) \leqslant \text{Max}_r(\mathcal{F})$.

We already know (see Lemma 1.34) that if a family $\mathcal{F}$ of feasible solutions is homogeneous (all sets of $\mathcal{F}$ have the same cardinality), then there is no difference between the tropical circuit complexities of the minimization and the maximization problems on $\mathcal{F}$, that is, then we have $\text{Min}_1(\mathcal{F}) = \text{Max}_1(\mathcal{F})$. This, however, is no longer the case for larger approximation factors $r > 1$.

Consider, for example, the family $\mathcal{F}$ of all perfect matchings in $K_{n,n}$, viewed as sets of their edges. We will show (see Corollary 4.3) that $\text{Min}_r(\mathcal{F}) = n^{\Omega(\log n)}$ holds for *any* finite approximation factor $r = r(n) \geqslant 1$. On the other hand, $\text{Max}_r(\mathcal{F}) \leqslant n^2$ holds for a (large but finite) factor $r = n$: just take a (max, +) circuit that outputs the maximum weight of a single edge.

Thus, unlike for *exactly solving* tropical circuits, the *approximation* complexities of minimization and maximization problems (on the same sets of feasible solutions) may be very different, that is, in the case of approximation, minimization, and maximization "divorce."

### 4.1.1  Boolean Lower Bound

As before, for a finite set $A \subseteq \mathbb{N}^n$ of vectors, let $\text{Bool}(A)$ denote the minimum size of a monotone Boolean $(\vee, \wedge)$ circuit computing the monotone Boolean function

$$\hat{f}(x) = \bigvee_{a \in A} \bigwedge_{i \in \text{sup}(a)} x_i$$

defined by the set $A$. By Lemma 1.30, we already know that $\text{Min}_r(A) \geqslant \text{Bool}(A)$ holds for every antichain $A \subseteq \{0, 1\}^n$ and any finite approximation factor $r = r(n) \geqslant 1$. We will now show that, actually, the same "Boolean bound" holds not only for (min, +) but even for (min, max, +) circuits, where *both* operations min and max are allowed. As we will later see (Theorem 6.5), (min, max, +) circuits solving minimization problems can be exponentially smaller than (min, +) circuits. Still, the following lemma shows that the Boolean lower bound holds also for (min, max, +) circuits.

**Lemma 4.1 (Jukna, Seiwert, and Sergeev [7])** *Let $A \subseteq \mathbb{N}^n$ be any finite set of vectors. Every* (min, max, +) *circuit approximating the minimization problem $f(x) = \min_{a \in A} \sum_{i=1}^{n} a_i x_i$ on $A$ within a finite factor $r = r(n) \geqslant 1$ must have at least $\text{Bool}(A)$ gates.*

**Proof** The lemma was proved in [7] only for factor $r = 1$, but the same argument works also for any factor $r \geqslant 1$. Take a (min, max, +) circuit $\Phi$ approximating the minimization problem $f$ on the set $A$ within a finite factor $r \geqslant 1$. The Boolean function defined by the set $A$ is $\hat{f}(x) = \bigvee_{a \in A} \bigwedge_{i \in \sup(a)} x_i$. Our goal is to show that, without increasing the total number of gates, the circuit $\Phi$ can be transformed into a monotone Boolean $(\vee, \wedge)$ circuit computing $\hat{f}$.

The *dual* of a Boolean function $h(x_1, \ldots, x_n)$ is $\neg h(\neg x_1, \ldots, \neg x_n)$. That is, we reverse the input bits as well as the output bit. In particular, due to $\neg(x \vee y) = \neg x \wedge \neg y$, $\neg(x \wedge y) = \neg x \vee \neg y$, and $\neg(\neg x) = x$, the dual of the Boolean function $\hat{f}(x) = \bigvee_{a \in A} \bigwedge_{i \in \sup(a)} x_i$ defined by $A$ is the Boolean function

$$g(x) = \bigwedge_{a \in A} \bigvee_{i \in \sup(a)} x_i .$$

The dual of a monotone Boolean circuit is obtained by interchanging AND and OR gates and input constants 0 and 1. It is well known and easy to verify (see, for example, [3, Theorem 1.3]) that duals of Boolean circuits compute the duals of the functions computed by the (original) circuits.

We now transform our (min, max, +) circuit $\Phi$ into monotone Boolean $(\vee, \wedge)$ circuit $\phi$ as follows: replace each positive input constant by 1, and do the following replacements of gates: $\min\{u, v\} \mapsto u \wedge v$, $\max\{u, v\} \mapsto u \vee v$, and $u + v \mapsto u \vee v$. Since Boolean circuits and their duals have the same size, Lemma 4.1 is a direct consequence of the following claim. $\square$

**Claim 4.2** *The Boolean circuit $\phi$ computes the dual $g$ of $\hat{f}$.*

To prove the claim, consider the following mapping $z \mapsto [z]$ from $\mathbb{R}_+$ to $\{0, 1\}$ by letting $[0] = 0$ and $[z] = 1$ for any $z > 0$. For a vector $x \in \mathbb{R}_+^n$, let $[x] = ([x_1], \ldots, [x_n]) \in \{0, 1\}^n$ denote the corresponding Boolean vector.

Since the (min, max, +) circuit $\Phi$ approximates our minimization problem $f(x) = \min_{a \in A} \langle a, x \rangle$ within a finite factor $r \geqslant 1$, we know that the inequalities $f(x) \leqslant \Phi(x) \leqslant r \cdot f(x)$ and, hence, also $[f(x)] \leqslant [\Phi(x)] \leqslant r \cdot [f(x)]$ hold for all $x \in \mathbb{R}_+^n$. Since the signum function $[x]$ takes only two values 0 or 1, this is equivalent to $[\Phi(x)] = [f(x)]$ for all $x \in \mathbb{R}_+^n$.

For the Boolean version $\phi$ of $\Phi$, we also have $[\Phi(x)] = \phi([x])$ for all $x \in \mathbb{R}_+^n$. Indeed, if $\Phi = x_i$ is a variable or $\Phi = c$ is a constant $c \in \mathbb{R}_+$, then the Boolean version $\phi$ of $\Phi$ is either a Boolean variable $[x_i]$ or a Boolean constant $[c] \in \{0, 1\}$. The rest follows by induction on the size of the circuit $\Phi$ using the equalities holding for all numbers $u, v \in \mathbb{R}_+$:

$$[\min\{u, v\}] = [u] \wedge [v] \text{ and } [\max\{u, v\}] = [u] \vee [v] = [u + v].$$

Thus, for all inputs $x \in \mathbb{R}_+^n$, we have $\phi([x]) = [\Phi(x)] = [f(x)]$, where

$$[f(x)] = \left[ \min_{a \in A} \sum_{i=1}^{n} a_i x_i \right] = \bigwedge_{a \in A} \left[ \sum_{i=1}^{n} a_i x_i \right] = \bigwedge_{a \in A} \bigvee_{i \in \sup(a)} [x_i] = g([x]),$$

that is, $\phi([x]) = g([x])$ holds for all $x \in \mathbb{R}_+^n$. In particular, $\phi(x) = g(x)$ holds for all $x \in \{0, 1\}^n$ (because then $[x] = x$), that is, the Boolean circuit $\phi$ computes the Boolean function $g$, as desired.                                                                         $\square$

Lemma 4.1 directly translates known lower bounds on the size of monotone Boolean circuits computing Boolean functions to the lower bounds on the size of tropical (min, +) and even (min, max, +) circuits approximating the corresponding minimization problems. We restrict ourselves with several examples.

The *n-assignment problem* is as follows: given an assignment of nonnegative real weights to the edges of the complete bipartite $n \times n$ graph, compute the minimum weight of a perfect matching in this graph. The corresponding family of feasible solutions is here the family of all perfect matchings, viewed as sets of their edges.

**Corollary 4.3** *Every* (min, max, +) *circuit approximating the n-assignment problem within any finite factor must have at least* $n^{\Omega(\log n)}$ *gates.*

**Proof** The Boolean function defined by the family of feasible solutions of the assignment problem is the Boolean permanent function $\mathrm{per}_n(x) = \bigvee_{\sigma \in S_n} \bigwedge_{i=1}^n x_{i,\sigma(i)}$, where $S_n$ is the set of all $n!$ permutations of $\{1, \ldots, n\}$. Razborov [12] has proved that $\mathrm{per}_n$ requires monotone Boolean circuits of size $n^{\Omega(\log n)}$.                                                                         $\square$

Let $n$ be a prime power and $1 \leqslant d \leqslant n$ be an integer. The *polynomial* $(n, d)$-*design* is the family $\mathcal{F}$ of all $|\mathcal{F}| = n^d$ $n$-element subsets $\{(a, p(a)): a \in GF(n)\}$ of the grid $GF(n) \times GF(n)$, where $p = p(x)$ ranges over all $n^d$ univariate polynomials of degree at most $d - 1$ over $GF(n)$.

**Corollary 4.4** *Let* $d \leqslant (1/2)\sqrt{n/\ln n}$. *The minimization problem on the polynomial* $(n, d)$-*design* $\mathcal{F}$ *can be solved by a* (min, +) *circuit of size* $2n^{d+1}$, *and every* (min, max, +) *circuit approximating this problem within any finite factor must have at least* $n^{\Omega(d)}$ *gates.*

**Proof** The upper bound is trivial: just compute the $|\mathcal{F}| = n^d$ sums and take their minimum. On the other hand, by improving an earlier bound of Andreev [2], Alon and Boppana [1] have shown that any monotone Boolean circuit computing the Boolean function $f(x) = \bigvee_{S \in \mathcal{F}} \bigwedge_{i \in S} x_i$ requires $n^{\Omega(d)}$ gates, as long as $d \leqslant (1/2)\sqrt{n/\ln n}$.                                                                         $\square$

To give an another application of Lemma 4.1, consider the tropical (min, +) version

$$\mathrm{MinS}_k^n(x_1, \ldots, x_n) = \min \left\{ \sum_{i \in S} x_i : S \subseteq [n], |S| = k \right\}$$

of the elementary symmetric polynomial $S_{n,k}(x) = \sum_{|S|=k} \prod_{i \in S} x_i$. A *formula* is a circuit with all gates of fanout 1.

**Corollary 4.5** *Let the threshold* $2 \leqslant k \leqslant n/2$ *be even. The problem* $\mathrm{MinS}_k^n$ *can be solved by a* (min, +) *circuit of size* $\mathcal{O}(kn)$, *but every* (min, max, +) *formula* approximating $\text{MinS}_k^n$ or $\text{MinS}_{n-k+1}^n$ *within any finite factor must use at least* $(1/2)nk\log(n/k)$ *gates.*

**Proof** The upper bound for (min, +) *circuits* is shown in Example 1.6. To show the lower bound for (min, max, +) *formulas*, recall that the threshold-$k$ function is a monotone Boolean function $\text{Th}_k^n(x_1,\ldots,x_n)$ which outputs 1 iff $x_1 + \cdots + x_n \geqslant k$. That is, $\text{Th}_k^n(x) = \bigvee_{|S|=k} \bigwedge_{i\in S} x_i$. Radhakrishnan [10] has shown that, for every $2 \leqslant k \leqslant n/2$, every monotone Boolean formula computing $\text{Th}_k^n(x)$ must use at least $(1/2)nk\log(n/k)$ gates. Since $\text{Th}_k^n$ is the Boolean version of the problem $\text{MinS}_k^n$, the desired lower bound on the size of (min, max, +) formulas approximating $\text{MinS}_k^n$ follows from Lemma 4.1.

To show that the same lower bound $(1/2)nk\log(n/k)$ also holds for (min, +) formulas approximating $\text{MinS}_{n-k+1}^n$, it is enough to observe that $\text{Th}_{n-k+1}^n$ is the dual of the function $\text{Th}_k^n$: an input vector $x \in \{0,1\}^n$ contains $\geqslant n - k + 1$ ones iff it contains $\leqslant k - 1$ zeros. □

**Remark 4.6** By an interesting adoption of the graph entropy lower bound argument of Newman and Wigderson [9] to tropical circuits, Mahajan, Nimbhorkar, and Tawari [8] have shown that the minimum number of leaves in a (min, +) formula computing $\text{MinS}_2^n$ is *exactly* $n\log n$. □

**Remark 4.7** Corollary 4.5 shows that tropical (min, +) *circuits* can be by a logarithmic factor smaller than (min, +) *formulas*. We will show in Sect. 5.5 that the gap between tropical circuits and tropical formulas can be even *super-polynomial* (see Remark 5.12). □

## 4.1.2  Boolean Upper Bound

We will now show that there is also a partial *converse* of Lemma 4.1: Boolean circuits can be used to prove also *upper* (not only lower) bounds on the size of approximating (min, max, +) and even of approximating (min, +) circuits.

Let $f : \{0,1\}^n \to \{0,1\}$ be a monotone Boolean function (i.e., $b \leqslant a$ and $f(b) = 1$ imply $f(a) = 1$). The *lowest one* of $f$ is a vector $a \in f^{-1}(1)$ such that $f(b) = 0$ for all $b \leqslant a$, $b \neq a$. Note that if $A$ is the set of lowest ones of $f$, then $A$ is an antichain and defines the function $f$, that is, $f(x) = \bigvee_{a\in A} \bigwedge_{i\in\sup(a)} x_i$.

Let $\phi$ be a monotone Boolean $(\vee, \wedge)$ circuit computing a monotone Boolean function $f : \{0,1\}^n \to \{0,1\}$, and let $B \subseteq \mathbb{N}^n$ be the set of exponent vectors produced by a monotone Boolean $(\vee, \wedge)$ circuit $\phi$ computing $f$. If $A \subseteq f^{-1}(1)$ is the set of lowest ones of $f$, then Lemma 1.15 implies that the set $B$ has the following two properties: $B \subseteq A^\uparrow$ and

(∗)  for every vector $a \in A$ there is a vector $b \in B$ with $\sup(b) = \sup(a)$.

We now restrict the magnitude of vectors $b$ in (∗) and say that the circuit $\phi$ is a *read-$r$ circuit* if for every vector $a \in A$ there is a vector $b \in B$ such that $\sup(b) =$

$\sup(a)$ and $b \leqslant r \cdot a$. In particular, a circuit $\phi$ is a *read*-1 circuit iff the inclusions $A \subseteq B \subseteq A^{\uparrow}$ hold.

**Remark 4.8** An equivalent and, apparently, more intuitive definition of read-$r$ circuits is using arithmetic circuits. Let $\phi$ be a monotone Boolean $(\vee, \wedge)$ circuit computing a monotone Boolean function $f$. The arithmetic $(+, \times)$ version of $\phi$ (obtained by replacing OR gates by $+$ gates and AND gates by $\times$ gates) produces (purely syntactically) some polynomial $P_{\phi}(x) = \sum_{S \in \mathcal{F}} c_S \prod_{i \in S} x_i^{d_i}$ (with integer coefficients $c_S \geqslant 1$ telling how often the corresponding monomial $\prod_{i \in S} x_i^{d_i}$ appears in the polynomial). By Lemma 1.15, the polynomial $P_{\phi}(x)$ has the following two properties:

(i)  For every monomial $\prod_{i \in S} x_i^{d_i}$ of $P_{\phi}$, the term $\bigwedge_{i \in S} x_i$ is an implicant[1] of $f$; this is the property $B \subseteq A^{\uparrow}$ in Lemma 1.15.
(ii) For every prime implicant $\bigwedge_{i \in I} x_i$ of $f$, there is a monomial $\prod_{i \in S} x_i^{d_i}$ in $P_{\phi}$ with $S = I$ and all $d_i \geqslant 1$; this is the property $(*)$.

The circuit $\phi$ is then a read-$r$ circuit, if the monomials $\prod_{i \in S} x_i^{d_i}$ of $P_{\phi}$ guaranteed by (ii) satisfy an additional condition $d_i \leqslant r$ for all $i \in I$. There are no restrictions on the degrees of other monomials of $P_{\phi}$. In particular, $\phi$ is automatically a read-$r$ circuit if the polynomial $P_{\phi}$ has no variable $x_i$ with degree $> r$ at all. $\qquad\qquad\square$

For a finite set $A \subseteq \mathbb{N}^n$, let

$$\texttt{Bool}_r(A) := \text{min size of a monotone read-}r\,(\vee, \wedge) \text{ circuit computing the}$$

$$\text{Boolean function } f(x) = \bigvee_{a \in A} \bigwedge_{i \in \sup(a)} x_i \text{ defined by } A.$$

The *tropical* (min, +) *version* of a Boolean $(\vee, \wedge)$ circuit is obtained by replacing $\vee$ gates by min gates and $\wedge$ gates by $+$ gates (here, we only consider constant-free Boolean circuits: the constant inputs 0 and 1 can be easily eliminated, as long as the computed Boolean function is non-constant). Similarly, the *Boolean version* of a constant-free (min, +) circuit is obtained by replacing min gates by $\vee$ gates and $+$ gates by $\wedge$ gates. That is, we replace the "addition" gates by "addition" gates and "multiplication" gates by "multiplication" gates of the corresponding semirings. Thus, the sets of "exponent" vectors produced by a circuit and by its version in another semiring are the *same*.

**Lemma 4.9** *Let $A \subseteq \{0, 1\}^n$. For every $r \geqslant 1$, we have $\texttt{Bool}(A) \leqslant \texttt{Min}_r(A) \leqslant \texttt{Bool}_r(A)$. For $r = 1$, we have $\texttt{Min}_1(A) = \texttt{Bool}_1(A)$.*

This is a weaker version of a tighter result proved in [6, Theorem 7.1].

**Proof** The inequality $\texttt{Bool}(A) \leqslant \texttt{Min}_r(A)$ follows from Lemma 4.1. To show the inequality $\texttt{Min}_r(A) \leqslant \texttt{Bool}_r(A)$, recall that the lower antichain $A'$ of $A$ consists of all vectors $a \in A$ such that $a \not\leqslant a'$ for all $a' \in A \setminus \{a\}$. Since both sets $A$ and

---

[1] A term $t = \bigwedge_{i \in S} x_i$ is an *implicant* of a monotone Boolean function $f$ if the inequality $t(a) \leqslant f(a)$ holds for all $a \in \{0, 1\}^n$ and is a *prime* implicant of $f$ if this no longer holds for any subterm $t' = \bigwedge_{i \in S \setminus \{j\}} x_i$ with $j \in S$.

$A'$ define the same monotone Boolean function, we have $\text{Bool}_r(A') = \text{Bool}_r(A)$. So, we can assume that the given set $A \subseteq \{0, 1\}^n$ is an *antichain*.

Let $\hat{f}(x) = \bigvee_{a \in A} \bigwedge_{i \in \sup(a)} x_i$ be the monotone Boolean function defined by the antichain $A$; hence, $A$ is exactly the set of lowest ones of $\hat{f}$. Let $\phi$ be a monotone Boolean read-$r$ circuit of size $\text{Bool}_r(A)$ computing $\hat{f}$, and let $B \subseteq \mathbb{N}^n$ be the set of exponent vectors produced by $\phi$. The tropical (min, +) version $\Phi$ of $\phi$ produces the same set $B$ of "exponent" vectors. Hence, the circuit $\Phi$ solves the minimization problem $f_B(x) = \min_{b \in B} \langle b, x \rangle$ on the set $B$. The minimization problem on the given set $A$ is $f_A(x) = \min_{a \in A} \langle a, x \rangle$. It remains to show that for every input weighting $x \in \mathbb{R}_+^n$, the inequalities $f_A(x) \leqslant f_B(x) \leqslant r \cdot f_A(x)$ hold. So, take an arbitrary input weighting $x \in \mathbb{R}_+^n$.

Since the Boolean circuit $\phi$ computes $\hat{f}$, Lemma 1.15 gives the inclusion $B \subseteq A^\uparrow$, that is, for every vector $b \in B$, there is a vector $a \in A$ such that $a \leqslant b$. This gives the first inequality $f_A(x) \leqslant f_B(x)$. To show the second inequality $f_B(x) \leqslant r \cdot f_A(x)$, take a vector $a \in A$ on which the minimum $f_A(x) = \langle a, x \rangle$ is achieved. Since $\phi$ is a read-$r$ circuit, there is a vector $b \in B$ such that $\sup(b) = \sup(a)$ and $b \leqslant r \cdot a$. This yields $f_B(x) \leqslant \langle b, x \rangle = \sum_{i \in \sup(b)} b_i x_i = \sum_{i \in \sup(a)} b_i x_i \leqslant \sum_{i \in \sup(a)} r a_i x_i = r \cdot \langle a, x \rangle = r \cdot f_A(x)$, as desired.

Now suppose that $r = 1$. It is enough to show the inequality $\text{Bool}_1(A) \leqslant \text{Min}_1(A)$. For this, take any (min, +) circuit $\Phi$ of size $\text{Min}_1(A)$ solving the minimization problem on the set $A$ of feasible solutions. By the antichain lemma (Lemma 1.33), we can assume that $A$ is an antichain (if not, then consider the lower antichain of $A$). By Lemma 1.29, we can also assume that the circuit $\Phi$ is constant-free. The Boolean $(\vee, \wedge)$ version $\phi$ of $\Phi$ produces the same set $B \subseteq \mathbb{N}^n$ of "exponent" vectors as the (min, +) circuit $\Phi$. Since the circuit $\Phi$ solved the minimization problem on $A$, Lemma 1.22 gives the inclusions $A \subseteq B \subseteq A^\uparrow$, meaning that the Boolean version $\phi$ of $\Phi$ is a read-1 circuit and, by Lemma 1.15, $\phi$ computes the Boolean function $\hat{f}$ defined by the set $A$ (inclusion $A \subseteq B$ is even stronger than the inclusion $\text{Sup}(A) \subseteq \text{Sup}(B)$ required by Lemma 1.15). □

Lemma 4.9 has an interesting consequence: one can design tropical (min, +) circuits approximating a given minimization problem within a given factor $r$ by designing monotone read-$r$ *Boolean* circuits solving the decision version of this problem. The following lemma illustrates how does this happen.

**Lemma 4.10** *Let $A \subseteq \{0, 1\}^{n \times n}$ be the set of Boolean $n \times n$ matrices $a = (a_{i,j})$ with at least one 1 in each row and in each column. Then $\text{Min}_1(A) = 2^{\Omega(n)}$ but $\text{Min}_2(A) \leqslant 2n^2$.*

**Proof** The lower bound $\text{Min}_1(A) = 2^{\Omega(n)}$ is given by Lemma 1.43. To show the upper bound $\text{Min}_2(A) \leqslant 2n^2$, let $f(x) = \bigvee_{a \in A} \bigwedge_{a_{i,j}=1} x_{i,j}$ be the monotone Boolean function defined by the set $A$. Hence, $f(x) = 1$ iff each row and each column of the matrix $x = (x_{i,j})$ has at least one 1. This function can be computed by a monotone Boolean circuit (in fact, a formula)

$$\phi(x) = \bigwedge_{i=1}^{n} \left( \bigvee_{j=1}^{n} x_{i,j} \right) \wedge \bigwedge_{j=1}^{n} \left( \bigvee_{i=1}^{n} x_{i,j} \right)$$

of size $2(n-1)^2 + 1 \leqslant 2n^2$. By Lemma 4.9, it is enough to show that this circuit $\phi$ is a read-2 circuit. The polynomial produced by the arithmetic $(+, \times)$ version of the circuit $\phi$ is of the form

$$P(x) = \prod_{i=1}^{n} \left( \sum_{j=1}^{n} x_{i,j} \right) \prod_{j=1}^{n} \left( \sum_{i=1}^{n} x_{i,j} \right).$$

Since no variable $x_{i,j}$ appears in this polynomial with degree larger than 2, the circuit $\phi$ is a read-2 circuit, as desired.                                                    □

## 4.2   Approximating (Max, +) Circuits

In Sect. 4.1.1, we have shown that the *minimization* problem on some families $\mathcal{F} \subseteq 2^{[n]}$ cannot be efficiently approximated by (min, +) circuits within *any* finite factor $r = r(n) \geqslant 1$. On the other hand, in the case of *maximization* problems, the approximation factor is always finite. Namely, we always have $\mathrm{Max}_r(\mathcal{F}) \leqslant n - 1$ for $r = n$: since the weights are nonnegative, we can just use the trivial (max, +) circuit computing $\max\{x_1, \ldots, x_n\}$ (under a natural assumption that every $i \in [n]$ belongs to at least one set $F \in \mathcal{F}$).

### 4.2.1   Counting Fails

There is an even more substantial difference between approximating (min, +) and (max, +) circuits than just the "bounded versus unbounded approximation factors" phenomenon: unlike for (min, +) circuits, even *counting* fails to yield nontrivial lower bounds on the size of approximating (max, +) circuits, and this happens already for very small approximation factors $r = 1 + o(1)$. The reason for this is that a single small (max, +) circuit can approximate within such a factor a huge (doubly exponential) number of distinct maximization problems.

To show this, consider the *top $k$-of-$n$ selection* problem $\mathrm{MaxS}_k^n$, which given nonnegative weights $x_1, \ldots, x_n$ outputs the sum of the $k$ largest weights:

$$\mathrm{MaxS}_k^n(x) = \max \left\{ \sum_{i \in S} x_i : S \subseteq [n], |S| = k \right\}.$$

This is the tropical (max, +) version of the elementary symmetric polynomial $S_{n,k}(x) = \sum_{|S|=k} \prod_{i \in S} x_i$. A (max, +) circuit $\Phi_k^n$ of size $\mathcal{O}(nk)$ solving the top $k$-of-$n$ selection problem $\mathrm{MaxS}_k^n$ (within factor $r = 1$) is constructed in Example 1.6.

Let (as before) $\binom{[n]}{m}$ denote the family of all $m$-element subsets of $[n] = \{1, \ldots, n\}$. Say that a family $\mathcal{F} \subseteq \binom{[n]}{m}$ is $k$-*dense* if every set $S \in \binom{[n]}{k}$ is contained in at least one set of $\mathcal{F}$.

**Proposition 4.11** *The* (max, +) *circuit* $\Phi_k^n$ *approximates the maximization problem on every $k$-dense family* $\mathcal{F} \subseteq \binom{[n]}{m}$ *within the factor $r = m/k$. In particular,* $\mathrm{Max}_r(\mathcal{F}) = \mathcal{O}(nk)$ *holds for every $k$-dense family $\mathcal{F}$.*

**Proof** Let $\mathcal{F} \subseteq \binom{[n]}{m}$ be a $k$-dense family. The maximization problem on $\mathcal{F}$ is $f(x) = \max_{F \in \mathcal{F}} \sum_{i \in F} x_i$. Since the weights are nonnegative, the $k$-denseness of $\mathcal{F}$ ensures that $f(x) \geqslant \mathrm{MaxS}_k^n(x)$. On the other hand, since no feasible solution $F \in \mathcal{F}$ of the problem $f$ has more than $m$ elements, the maximum weight $f(x)$ of a feasible solution cannot exceed $m/k$ times the sum of weights of $k$ heaviest elements in this solution. Hence, $f(x) \leqslant (m/k) \cdot \mathrm{MaxS}_k^n(x)$, as desired. $\square$

We will now show that the single (max, +) circuit $\Phi_k^n$ approximates within a small factor $r = 1 + o(1)$ a huge number of (distinct) maximization problems. Let $n = 2m$ for $m \geqslant 2$.

**Lemma 4.12** *At least $2^{\binom{n}{m}/n}$ families $\mathcal{F} \subseteq \binom{[n]}{m}$ are $k$-dense for $k = m - 1$.*

**Proof** The Hamming distance between two sets $A$ and $B$ is $\mathrm{dist}(A, B) = |A \setminus B| + |B \setminus A|$. A family $\mathcal{H}$ is a *Hamming family* if $\mathrm{dist}(A, B) > 2$ holds for all $A \neq B \in \mathcal{H}$.

An easy argument of Graham and Sloane [4] shows that a Hamming family $\mathcal{H} \subseteq \binom{[n]}{m}$ with $|\mathcal{H}| \geqslant \frac{1}{n}\binom{n}{m}$ sets exists. For $l = 0, 1, \ldots, n - 1$, let $\mathcal{H}_l \subseteq \binom{[n]}{m}$ consist of all $m$-element sets $S \subseteq [n]$ such that $\sum_{i \in S}(i - 1) \equiv l \pmod{n}$. Since each set of $\binom{[n]}{m}$ belongs to at least one of the $n$ families $\mathcal{H}_0, \mathcal{H}_1, \ldots, \mathcal{H}_{n-1}$, it is enough to show that each $\mathcal{H}_l$ is a Hamming family. If $\mathrm{dist}(S, T) = 2$ held for some two sets $S \neq T$ of $\mathcal{H}_l$, then $S = A \cup \{s\}$ and $T = A \cup \{t\}$ would hold for some $(m - 1)$-element set $A \subseteq [n]$ and distinct numbers $s \neq t$ in $\{0, 1, \ldots, n-1\} \setminus A$. But then we would have $a + s \equiv l \pmod{n}$ and $a + t \equiv l \pmod{n}$ for $a = \sum_{i \in A}(i - 1)$. This would imply $s \equiv t \pmod{n}$, which is impossible because $s \neq t$ and both numbers $s$ and $t$ are at most $n - 1$.

Thus, since sub-families of Hamming families are also Hamming families, there are at least $2^{\binom{n}{m}/n}$ Hamming families $\mathcal{H} \subseteq \binom{[n]}{m}$. It remains to show that the uniform complement $\mathcal{H}^c = \binom{[n]}{m} \setminus \mathcal{H}$ of every Hamming family $\mathcal{H} \subseteq \binom{[n]}{m}$ is $(m - 1)$-dense. To show this, take any set $T \in \binom{[n]}{m-1}$ and two distinct elements $a \neq b$ outside $T$, and consider the $m$-element sets $A = T \cup \{a\}$ and $B = T \cup \{b\}$. Since the Hamming distance between $A$ and $B$ is 2, they cannot *both* belong to the family $\mathcal{H}$. So, at least one of them must belong to $\mathcal{H}^c$, as desired. $\square$

For a family $\mathcal{F} \subseteq 2^{[n]}$, let $\mathrm{Bool}(\mathcal{F})$ denote the minimum size of a monotone Boolean circuit computing the Boolean function $f(x) = \bigvee_{S \in \mathcal{F}} \bigwedge_{i \in S} x_i$.

**Corollary 4.13** *Let* $m \geqslant 2$, $n = 2m$, *and* $k = m - 1$. *There are* $2^{2^{\Omega(n)}}$ *families* $\mathcal{F} \subseteq \binom{[n]}{m}$ *such that* $\mathtt{Max}_1(\mathcal{F}) = 2^{\Omega(n)}$ *and* $\mathtt{Bool}(\mathcal{F}) = 2^{\Omega(n)}$, *but* $\mathtt{Max}_{1+o(1)}(\mathcal{F}) = \mathcal{O}(n^2)$ *holds for each of these families.*

Moreover, the maximization problem on each of these families $\mathcal{F}$ is approximated within a small factor $r = 1 + o(1)$ by *one and the same* (max, +) circuit $\Phi_k^n$ of size $\mathcal{O}(nk) = \mathcal{O}(n^2)$. In particular, and unlike for approximating (min, +) circuits, there is no "Boolean bound" (analogue of Lemma 4.1) for approximating (max, +) circuits: the gap $\mathtt{Bool}(\mathcal{F})/\mathtt{Max}_r(\mathcal{F})$ can be exponential even for very small approximation factors $r = 1 + o(1)$.

*Proof of Corollary 4.13* Lemma 4.12 gives us $M \geqslant 2^{\binom{n}{m}/n} \geqslant 2^{2^n/2n^{3/2}}$ distinct families $\mathcal{F}_1, \ldots, \mathcal{F}_M \subseteq \binom{[n]}{m}$, each of which is $k$-dense for $k = m - 1$. By Proposition 4.11, the (max, +) circuit $\Phi_k^n$ solving the top $k$-of-$n$ selection problem $\mathtt{MaxS}_k^n$ approximates the maximization problem $f_i(x) = \max\{\sum_{i \in S} x_i : S \in \mathcal{F}_i\}$ on each of these families $\mathcal{F}_i$ within the factor $r = m/k \leqslant 1 + 2/n = 1 + o(1)$.

On the other hand, the maximization problems $f_1, \ldots, f_M$ on all families $\mathcal{F}_1, \ldots, \mathcal{F}_M$ are *distinct* (as functions). Assume for a contradiction that there are $i \neq j$ such that $f_i(x) = f_j(x)$ holds for all $x \in \mathbb{R}_+^n$. Then, by Lemma 1.24, we have the inclusions $\mathcal{F}_i \subseteq \mathcal{F}_j \subseteq \mathcal{F}_i^{\downarrow}$, where $\mathcal{F}^{\downarrow} := \{S \subseteq [n] : S \subseteq F \text{ for some } F \in \mathcal{F}\}$. Since sets in $\mathcal{F}_i$ and $\mathcal{F}_j$ have the same cardinality, $\mathcal{F}_j \subseteq \mathcal{F}_i^{\downarrow}$ implies $\mathcal{F}_j \subseteq \mathcal{F}_i$ and, hence, also $\mathcal{F}_i = \mathcal{F}_j$, a contradiction (all families $\mathcal{F}_i$ are distinct).

Let $\Phi_1, \ldots, \Phi_M$ be tropical (max, +) circuit solving the maximization problems $f_1, \ldots, f_M$. By Lemma 1.29, we can assume that all these circuits are constant-free. By Proposition 1.1, there are at most $L(n, s) = 2^{ct \log(s+n)}$ distinct constant-free (max, +) circuits $\Phi(x_1, \ldots, x_n)$ of size $\leqslant s$, where $c$ is a constant. For $s := 2^n/n^3$, we have $\log L(n, s) \leqslant c2^n/n^2 \ll 2^n/2n^{3/2} = \log M$. Thus, since one single (max, +) circuit can solve (exactly) only one maximization problem, all but a neglected portion of circuits $\Phi_1, \ldots, \Phi_M$ must have more than $2^n/n^3$ gates.

That all but a neglected portion of Boolean functions $\bigvee_{S \in \mathcal{F}_i} \bigwedge_{j \in S} x_j$, $i = 1, \ldots, M$, require monotone Boolean $(\vee, \wedge)$, and even non-monotone $(\vee, \wedge, \neg)$ circuits with more than $2^n/n^3$ gates follow by the same counting argument. $\square$

**Remark 4.14 (Matroids)** A uniform family $\mathcal{F} \subseteq \binom{[n]}{m}$ is a *matroid* (or, more precisely, a family of bases of a matroid) if the *basis exchange axiom* holds: if $A \neq B \in \mathcal{F}$, then for every $a \in A \setminus B$ there is a $b \in B \setminus A$ such that the set $(A \setminus \{a\}) \cup \{b\}$ belongs to $\mathcal{F}$. It can be shown that the families $\mathcal{F}$ guaranteed by Lemma 4.12 are, in fact, matroids. Thus, by a classical result of Rado [11], the maximization problem on each of these families $\mathcal{F}$ can be solved by the *greedy algorithm*: start with the empty solution and iteratively include the heaviest remaining element for which the extended solution remains feasible, i.e., is contained in at least one set of $\mathcal{F}$.

Let $\mathcal{H} \subseteq \binom{[n]}{m}$ be a Hamming family. To show that $\mathcal{H}^c$ is a matroid, assume the opposite. Since the family $\mathcal{H}^c$ is uniform, there must be two sets $A \neq B \in \mathcal{H}^c$ violating the basis exchange axiom: there is an $a \in A \setminus B$ such that $(A \setminus \{a\}) \cup \{b\} \notin \mathcal{H}^c$ for all $b \in B$. Observe that $B \setminus A$ must have at least two elements: held $B \setminus A = \{b\}$, then since both $A$ and $B$ have the same cardinality, the set $(A \setminus \{a\}) \cup \{b\}$ would coincide with $B$ and, hence, would belong to $\mathcal{H}^c$. So, take $b \neq c \in B \setminus A$ and consider the sets $S = (A \setminus \{a\}) \cup \{b\}$ and $T = (A \setminus \{a\}) \cup \{c\}$. Since the basis exchange axiom fails for $A$ and $B$, neither $S$ nor $T$ can belong to $\mathcal{H}^c$; hence, *both* sets $S$ and $T$ belong to the family $\binom{[n]}{m} \setminus \mathcal{H}^c = \mathcal{H}$. But $\mathrm{dist}(S, T) = |\{b, c\}| = 2$, a contradiction with $\mathcal{H}$ being a Hamming family. $\square$

It can also be shown that, with a large probability, the circuit $\Phi_k^n$ for the top $k$-of-$n$ selection problem approximates the maximization problem on random families of feasible solutions. A *random $m$-uniform family* $\mathcal{F} \subseteq \binom{[n]}{m}$ is obtained by including in $\mathcal{F}$ each $m$-element subset of $[n]$ independently with probability $1/2$.

**Proposition 4.15** *Let $n$ be a sufficiently large even integer, $m = n/2$, and $k = m - 2$. With probability $1 - o(1)$, the (max, +) circuit $\Phi_k^n$ approximates the maximization problem on a random $m$-uniform family $\mathcal{F} \subseteq \binom{[n]}{m}$ within the factor $r = 1 + o(1)$.*

**Proof** Since each $k$-element set is contained in $l := \binom{n-k}{2} = \Omega(n^2)$ sets of $\binom{[n]}{m}$, the probability that a fixed $k$-element set will be contained in *none* of the sets of $\mathcal{F}$ is $(1/2)^l = 2^{-\Omega(n^2)}$. So, by the union bound, the family $\mathcal{F}$ is *not* $k$-dense with probability at most $\binom{n}{m} \cdot 2^{-\Omega(n^2)} = 2^{-\Omega(n^2)}$. That is, the family $\mathcal{F}$ is $k$-dense with probability at least $1 - 2^{-\Omega(n^2)}$. By Proposition 4.11, with this probability, the circuit $\Phi_k^n$ approximates the maximization problem on a random family $\mathcal{F}$ within the factor $r = m/k = 1 + \mathcal{O}(1/n)$. □

### 4.2.2 Locally Balanced Decompositions

Decompositions of sets $B \subseteq \mathbb{N}^n$ of vectors into "balanced" rectangles $X + Y \subseteq B$ presented in Sect. 3.1 are "global" in a sense that the "balance measure" of a rectangle $X + Y$ in a decomposition is the measure $\mu(X)$ of the *entire* set $X$. As we have demonstrated in Chap. 3, such decompositions are useful when lower bounding the size of *exactly solving* tropical circuits (within the factor $r = 1$).

However, as Corollary 4.13 and Proposition 4.15 show, the situation with *approximating* (max, +) circuits (already for very small factors $r > 1$) is more delicate. This is why we will now go deeper into the structure of the produced sets of "exponent" vectors and consider "local" decompositions, where balancedness applies to *individual* vectors in the parts $X$ and $Y$ of rectangles $X + Y$.

A *norm measure* of vectors is an assignment $\mu : \mathbb{N}^n \to \mathbb{R}_+$ of nonnegative real numbers to vectors in $\mathbb{N}^n$ such that every 0-1 vector with at most one 1 gets norm at most 1, and the norm is monotone and subadditive in that $\mu(x) \leqslant \mu(x + y) \leqslant \mu(x) + \mu(y)$ holds for all vectors $x, y \in \mathbb{N}^n$. Examples of norm measures are $\mu(x) := x_1 + \cdots + x_n$ (the degree of vector $x$), $\mu(x) := |\sup(x)|$ (the number of nonzero positions in $x$), and $\mu(x) := \langle a, x \rangle$ for a fixed vector $a \in \{0, 1\}^n$.

A *locally balanced decomposition* of a finite set $B \subseteq \mathbb{N}^n$ of vectors is a collection of (not necessarily disjoint) rectangles $X + Y \subseteq B$ with the following property:

(∗) For every norm measure $\mu : \mathbb{N}^n \to \mathbb{R}_+$, for every vector $b \in B$ of norm $\mu(b) > 1$, and for every real number $\gamma$ satisfying $1/\mu(b) \leqslant \gamma < 1$, at least one of the rectangles $X + Y$ contains vectors $x \in X$ and $y \in Y$ such that $x + y = b$ and $\frac{\gamma}{2} < \mu(x)/\mu(b) \leqslant \gamma$.

Note that one can take different norm measures $\mu$ for different vectors $b \in B$; this "flexibility" will be important in the proof of a general lower bound for approximating (max, +) circuits (Lemma 4.19 and its set-theoretic version Lemma 4.20). Recall that $L(B)$ is the minimum size of a Minkowski $(\cup, +)$ circuit producing the set $B$.

**Lemma 4.16 (Jukna [5])** *Every finite set $B \subseteq \mathbb{N}^n$ of vectors has a locally balanced decomposition into at most $L(B)$ rectangles.*

***Proof*** Let $B \subseteq \mathbb{N}^n$ be a finite set, and let $\Phi$ be a Minkowski circuit of size $t = L(B)$ producing the set $B$. Since we have only $t$ gates in the circuit, it is enough to show that the collection of rectangles[2] $X_v + Y_v$ associated with the gates $v$ of $\Phi$ has the desired property $(*)$. So, fix any norm measure $\mu : \mathbb{N}^n \to \mathbb{R}_+$, any vector $b \in B$ of norm $m := \mu(b) > 1$, and any real number $1/m \leqslant \gamma < 1$. Our goal is to show that at least one of the rectangles $X_v + Y_v$ contains vectors $x \in X_v$ and $y \in Y_v$ such that $x + y = b$ and $\gamma/2 < \mu(x)/m \leqslant \gamma$ hold.

By a *decomposition* of the vector $b \in B$ at a gate $v$, we will mean a pair of vectors $x \in X_v$ and $y \in Y_v$ such that $b = x + y$. Note that at the output gate, we have the unique decomposition $b = x + \vec{0}$ of $b$ with $\mu(x) = \mu(b) = m$. $\qquad\square$

**Claim 4.17** *If vector $b$ has a decomposition $b = x + y$ at a gate $v$, then $b$ has a decomposition $b = x' + y'$ with $\frac{1}{2} \cdot \mu(x) \leqslant \mu(x') \leqslant \mu(x)$ at one of the two gates entering $v$.*

***Proof*** If $v = u \cup w$ is a union gate, then $X_v = X_u \cup X_w$ and, hence, $Y_v = Y_u \cap Y_w$. So, $b = x + y$ is a decomposition of $b$ at the gate $u$ (if $x \in X_u$) or at the gate $w$ (if $x \in X_w$), and the claim is trivial in this case with $\mu(x') = \mu(x)$.

Now let $v = u + w$ be a Minkowski sum gate, and $b = x + y$ for some $x \in X_v$ and $y \in Y_v$. In this case, the vector $x$ is of the form $x = x_u + x_w$ for some vectors $x_u \in X_u$ and $x_w \in X_w$. We can assume that $\mu(x_u) \geqslant \mu(x_w)$. Since $y \in Y_v$, we have $(X_u + X_w) + y \subseteq B$ and, hence, also $X_u + y' \subseteq B$ for the vector $y' := x_w + y$. So, the vector $y' := x_w + y$ belongs to the residue $Y_u$ of $B$ at gate $u$, implying that $b = x_u + y'$ is a decomposition of our vector $b$ at the gate $u$. The monotonicity of $\mu$ yields $\mu(x_u) \leqslant \mu(x_u + x_w) = \mu(x)$, and the subadditivity of $\mu$ yields $\mu(x) \leqslant \mu(x_u) + \mu(x_w) \leqslant 2 \cdot \mu(x_u)$; hence, $\frac{1}{2}\mu(x) \leqslant \mu(x_u) \leqslant \mu(x)$, as desired. $\qquad\square$

We now start at the output gate with the unique decomposition $b = b + \vec{0}$ of our vector $b$ and traverse an output-to-input path $\pi$ in the circuit by going *backward* and using the following rule:

o If $v$ is a currently reached gate and $b = x + y$ is a decomposition of $b$ at this gate, then go to that of the two inputs of $v$ at which vector $b$ has a decomposition $b = x' + y'$ with $\mu(x') \geqslant \frac{1}{2} \cdot \mu(x)$; if both input gates have this property, then go to any of them.

---

[2] Recall (from Sect. 2.4) that $X_v \subseteq \mathbb{N}^n$ is the set of vectors produced at the gate $v$, and $Y_v := \{y \in \mathbb{N}^n : X_v + y \subseteq B\}$ is the residue of the set $B$ at gate $v$.

Claim 4.17 ensures that we will eventually reach some input gate. At this gate, vector $b$ has a decomposition $b = x + y$, where either $x = \vec{0}$ (and $y = b$) or $x = \vec{e}_i$ (and $y = b - \vec{e}_i$). In both cases, we have that $\mu(x) \leqslant 1$, which is at most $\gamma m$, because $\gamma \geqslant 1/m$; recall that $m = \mu(b)$ is the norm of vector $b$. On the other hand, the decomposition $b = x + y$ with $x = b$ and $y = \vec{0}$ of the vector $b$ at the output gate has $\mu(x) = \mu(b) = m$, which is strictly larger than $\gamma m$, because $\gamma < 1$. So, there must be an edge $(u, v)$ in the path $\pi$ at which the jump from $\leqslant \gamma m$ to $> \gamma m$ happens. That is, there must be a decomposition $b = x + y$ of our vector $b$ at the gate $v$ and its decomposition $b = x' + y'$ at the gate $u$ such that $\mu(x) > \gamma m$ but $\mu(x') \leqslant \gamma m$. By Claim 4.17, we have $\mu(x') \geqslant \frac{1}{2} \cdot \mu(x)$. We have thus found a rectangle $X_u + Y_u$ and vectors $x' \in X_u$ and $y' \in Y_u$ such that $b = x' + y'$ and $\frac{1}{2}\gamma m < \mu(x') \leqslant \gamma m$, as desired. $\qquad\square$

**Remark 4.18** Note that by taking the norm measure $\mu(x) := x_1 + \cdots + x_n$, Lemma 4.16 directly yields the Hyafil–Valiant decomposition lemma (Lemma 3.3): if $X + Y \subseteq B$ for a homogeneous set $B \subseteq \mathbb{N}^n$, then both sets $X$ and $Y$ must be homogeneous. $\qquad\square$

### 4.2.3   Rectangle Bound for Approximating (Max, +) Circuits

Let $A \subseteq \{0, 1\}^n$, $r \geqslant 1$, and $0 < \gamma < 1$. We say that a vector $a \in A$ with $|a| := \langle a, a \rangle$ ones *appears* $(r, \gamma)$-*balanced* in a rectangle $X + Y$ if there are vectors $x \in X$ and $y \in Y$ such that

$$\langle a, x + y \rangle \geqslant |a|/r \quad \text{and} \quad \frac{\gamma}{2} < \frac{\langle a, x \rangle}{\langle a, x + y \rangle} \leqslant \gamma . \tag{4.1}$$

Recall that a set $B \subseteq \mathbb{N}^n$ *lies below* a set $A \subseteq \mathbb{N}^n$ if $\forall b \in B \, \exists a \in A$ such that $b \leqslant a$.

**Lemma 4.19 (Jukna and Seiwert [6])** *Let* $A \subseteq \{0, 1\}^n$, $r \geqslant 1$, *and* $0 < \gamma < 1$. *If* $\mathsf{Max}_r(A) \leqslant t$, *then there are $t$ or fewer rectangles $X + Y$ lying below $A$ such that every vector $a \in A$ with $|a| \geqslant r/\gamma$ ones appears $(r, \gamma)$-balanced in at least one of these rectangles.*

**Proof** Take a (max, +) circuit $\Phi$ of size $t = \mathsf{Max}_r(A)$ approximating the maximization problem $f(x) = \max_{a \in A} \langle a, x \rangle$ on $A$ within the factor $r$. By Lemma 1.29, we can assume that the circuit is constant-free. Let $B \subseteq \mathbb{N}^n$ be the set of vectors produced by $\Phi$. By Lemma 1.24, we know that the set $B$ has the following two properties:

(i)  For every vector $b \in B$, there is a vector $a \in A$ such that $b \leqslant a$.
(ii) For every vector $a \in A$, there is a vector $b \in B$ such that $\langle a, b \rangle \geqslant |a|/r$.

The Minkowski $(\cup, +)$ version $\Phi'$ of the circuit $\Phi$ (obtained by replacing max gates $\max\{u, v\}$ with union gates $u \cup v$ and addition gates $u + v$ with Minkowski sum gates) has the same size $t$ and produces the same set $B$; thus, $\mathrm{L}(B) \leqslant t$ holds. When applied to the Minkowski circuit $\Phi'$, Lemma 4.16 gives us a collection of at most $t$ rectangles $X + Y \subseteq B$ with the following property holding for every norm measure $\mu : \mathbb{N}^n \to \mathbb{R}_+$, for every vector $b \in B$ of norm $\mu(b) > 1$, and for every real number $\gamma$ satisfying $1/\mu(b) \leqslant \gamma < 1$:

($*$) For at least one of the rectangles $X + Y$, there are vectors $x \in X$ and $y \in Y$ such that $x + y = b$ and $\gamma/2 < \mu(x)/\mu(b) \leqslant \gamma$.

To show the desired property (4.1), fix an arbitrary vector $a \in A$ such that $|a| \geqslant r/\gamma$. Property (ii) of the set $B$ suggests to associate with $a$ the norm measure $\mu : \mathbb{N}^n \to \mathbb{R}_+$ defined by $\mu(x) := \langle a, x \rangle$. This is a norm measure because every 0-1 vector $x$ with at most one 1 gets norm $\mu(x) \leqslant 1$, and $\mu(x) \leqslant \mu(x + y) = \mu(x) + \mu(y)$ holds for all vectors $x, y \in \mathbb{N}^n$. Then, by (ii), there is a vector $b \in B$ of norm $\mu(b) = \langle a, b \rangle \geqslant |a|/r \geqslant 1/\gamma$. Since $\gamma < 1$, we have $1/\mu(b) \leqslant \gamma < 1$. Thus, by the property ($*$), at least one of our rectangles $X + Y$ contains vectors $x \in X$ and $y \in Y$ such that $x + y = b$ and

$$\frac{\gamma}{2} \leqslant \frac{\mu(x)}{\mu(b)} = \frac{\langle a, x \rangle}{\langle a, x + y \rangle} \leqslant \gamma \,.$$

Since $\langle a, x + y \rangle = \langle a, b \rangle \geqslant |a|/r$, the vectors $x \in X$ and $y \in Y$ have the desired properties (4.1). $\qquad\square$

### 4.2.4 Application: Approximation on Designs

In applications, it is convenient to translate Lemma 4.19 from the language of vectors to the language of sets. In the set-theoretic language, a *rectangle* is a family of sets specified by a pair $\mathcal{A}, \mathcal{B}$ of families that are *cross-disjoint* in that $A \cap B = \emptyset$ holds for all $A \in \mathcal{A}$ and $B \in \mathcal{B}$. The rectangle $\mathcal{R} = \mathcal{A} \vee \mathcal{B}$ itself consists of all sets $A \cup B$ with $A \in \mathcal{A}$ and $B \in \mathcal{B}$. The rectangle $\mathcal{R}$ lies *below* a family $\mathcal{F}$ if every set of $\mathcal{R}$ is contained in at least one set of $\mathcal{F}$.

Let $\mathcal{F}$ be a family of sets. Given an approximation factor $r \geqslant 1$ and a balance parameter $0 < \gamma < 1$, say that a set $F \in \mathcal{F}$ *appears* $(r, \gamma)$-*balanced* in a rectangle $\mathcal{R} = \mathcal{A} \vee \mathcal{B}$ if there are sets $A \in \mathcal{A}$ and $B \in \mathcal{B}$ such that

$$|F \cap (A \cup B)| \geqslant \frac{1}{r} \cdot |F| \quad \text{and} \quad \frac{\gamma}{2} \leqslant \frac{|F \cap A|}{|F \cap (A \cup B)|} \leqslant \gamma \,. \tag{4.2}$$

In particular, the set $F$ must share many elements with both sets $A$ and $B$ in common:

$$|F \cap A| \geqslant \frac{\gamma}{2r} \cdot |F| \quad \text{and} \quad |F \cap B| \geqslant \frac{1 - \gamma}{r} \cdot |F| \,.$$

The first inequality follows directly from the lower bound on $|F \cap A|$ in Eq. (4.2), while the second inequality follows from the upper bound on $|F \cap A|$ in Eq. (4.2):
$|F \cap B| \geqslant |F \cap (A \cup B)| - |F \cap A| \geqslant |F \cap (A \cup B)| - \gamma \cdot |F \cap (A \cup B)| = (1 - \gamma)|F \cap (A \cup B)| \geqslant \frac{1-\gamma}{r} \cdot |F|$.

By considering the sets $A \subseteq \{0, 1\}^n$ of characteristic vectors of sets in families $\mathcal{F} \subseteq 2^{[n]}$, Lemma 4.19 directly yields the following lower bound on the size of approximating (max, +) circuits.

**Lemma 4.20 (Set-Theoretic Version of the Rectangle Bound)** *Let* $r \geqslant 1$, $0 < \gamma < 1$, *and let* $\mathcal{F} \subseteq 2^{[n]}$ *be a family of sets, each with at least* $r/\gamma$ *elements. Suppose that for every rectangle* $\mathcal{R}$ *lying below* $\mathcal{F}$, *at most* $h$ *sets* $F \in \mathcal{F}$ *appear* $(r, \gamma)$-*balanced in* $\mathcal{R}$. *Then*

$$\mathsf{Max}_r(\mathcal{F}) \geqslant |\mathcal{F}|/h .$$

Thus, given an approximation factor $r \geqslant 1$, we can associate with every rectangle $\mathcal{R} = \mathcal{A} \vee \mathcal{B}$ the family $\mathcal{F}_\mathcal{R} \subseteq \mathcal{F}$ of all sets $F \in \mathcal{F}$ satisfying (4.2) for some sets $A \in \mathcal{A}$ and $B \in \mathcal{B}$. Then $\mathsf{Max}_r(\mathcal{F}) \geqslant |\mathcal{F}|/ \max |\mathcal{F}_\mathcal{R}|$, where the maximum is over all rectangles $\mathcal{R}$ lying below $\mathcal{F}$.

We already know (Corollary 4.13) that there *exist* many maximization problems for which a slight decrease of the allowed approximation factor from $r = 1 + o(1)$ to $r = 1$ can exponentially increase the size of (max, +) circuits. Using the rectangle bound (Lemma 4.20), one can show that such jumps can happen for arbitrarily *large* approximation factors $r$: a small decrease of the allowed approximation factor $r$ can make tractable problems intractable. Such jumps can be demonstrated on maximization problems whose families $\mathcal{F}$ of feasible solutions are "combinatorial designs."

An $(n, d)$-*design* is a family $\mathcal{F}$ of $n$-element sets which is $d$-*disjoint* in the sense that $|F \cap H| \leqslant d - 1$ holds for all distinct sets $F \neq H \in \mathcal{F}$. For a real number $l \geqslant 0$, let

$$\#_l(\mathcal{F}) := \max_{|S| \geqslant l} |\{F \in \mathcal{F}: F \supseteq S\}|$$

denote the maximal possible number of sets in $\mathcal{F}$ containing a fixed set with $l$ or more elements. For example, $\#_1(\mathcal{F}) = 1$ iff all sets of $\mathcal{F}$ are disjoint, and $\#_d(\mathcal{F}) = 1$ if $\mathcal{F}$ is an $(n, d)$-design.

**Lemma 4.21** *Let* $\mathcal{F}$ *be an* $(n, d)$-*design with* $1 \leqslant d \leqslant n/2$. *For the factor* $r = n/2d$, *we have* $\mathsf{Max}_r(\mathcal{F}) \geqslant |\mathcal{F}|/\#_l(\mathcal{F})$ *with* $l = d/2$.

The condition $d \leqslant n/2$ is only to ensure that the approximation factor $r$ is at least 1 (as it should be).

**Proof** We are going to apply the rectangle bound (Lemma 4.20) with the balance parameter $\gamma := 1/2$. Since $d \geqslant 1$, every set $F \in \mathcal{F}$ has $|F| = n \geqslant n/d = r/\gamma$ elements, as required by Lemma 4.20. Take an arbitrary rectangle $\mathcal{R} = \mathcal{A} \vee \mathcal{B}$ lying below $\mathcal{F}$, and let $\mathcal{F}_{\mathcal{R}} \subseteq \mathcal{F}$ be the family of all sets $F \in \mathcal{F}$ that appear $(r, \gamma)$-balanced in the rectangle $\mathcal{R}$. By Lemma 4.20, it is enough to show that $|\mathcal{F}_{\mathcal{R}}| \leqslant \#_l(\mathcal{F})$ holds.

Since every set of $\mathcal{F}_{\mathcal{R}}$ appears $(r, \gamma)$-balanced in the rectangle $\mathcal{R}$, for every $F \in \mathcal{F}_{\mathcal{R}}$ there are sets $A \in \mathcal{A}$ and $B \in \mathcal{B}$ such that

$$|F \cap A| \geqslant \frac{\gamma}{2r} \cdot |F| = \frac{\gamma}{2r} \cdot n = \frac{d}{2} = l \text{ and } |F \cap B| \geqslant \frac{1-\gamma}{r} \cdot n = d. \quad (4.3)$$

(We also know that $|F \cap (A \cup B)| \geqslant |F|/r$ holds, but we will not need this inequality in the current proof.) By Lemma 4.20, it is enough to show that $|\mathcal{F}_{\mathcal{R}}| \leqslant \#_l(\mathcal{F})$ holds.

Let $S$ be the union of all sets in $\mathcal{A}$. We can assume that $|A| \geqslant l$ holds for all sets $A \in \mathcal{A}$: smaller sets $A$ can be safely removed because they cannot fulfill $|F \cap A| \geq l$. So, $|S| \geqslant l$ holds as well, and it remains to show that

all sets of $\mathcal{F}_{\mathcal{R}}$ contain the set $S$ and, hence, $|\mathcal{F}_{\mathcal{R}}| \leqslant \#_{|S|}(\mathcal{F}) \leqslant \#_l(\mathcal{F})$.

To show this, take an arbitrary set $F \in \mathcal{F}_{\mathcal{R}}$. By Eq. (4.3), $|F \cap B_0| \geqslant d$ holds for some set $B_0 \in \mathcal{B}$. On the other hand, since the rectangle $\mathcal{R}$ lies below $\mathcal{F}$, every set $A \cup B_0$ with $A \in \mathcal{A}$ must lie in some set of $\mathcal{F}$. Since all these sets of $\mathcal{F}$ contain the set $B_0$ with $|B_0| \geqslant d$ elements and since the family $\mathcal{F}$ is $d$-disjoint, this implies that $all$ sets of $\mathcal{A} \vee \{B_0\}$ and, hence, also the set $S \cup B_0$ must be contained in $one$ set $F_0$ of $\mathcal{F}$. Since both sets $F$ and $F_0$ of $\mathcal{F}$ contain the same set $F \cap B_0$ of size $|F \cap B_0| \geqslant d$ and since the family $\mathcal{F}$ is $d$-disjoint, the equality $F = F_0$, and hence, the desired inclusion $S \subseteq F$ follows.                                                        □

A standard construction of large designs is by using polynomials. Let $n$ be a prime power and $1 \leqslant d \leqslant n$ be an integer. The $polynomial$ $(n, d)$-$design$ is the family $\mathcal{F}$ of all $|\mathcal{F}| = n^d$ $n$-element subsets $S_p = \{(a, p(a)) : a \in GF(n)\}$ of the grid $GF(n) \times GF(n)$, where $p = p(z)$ ranges over all $n^d$ univariate polynomials of degree at most $d - 1$ over $GF(n)$. To show that $\mathcal{F}$ is an $(n, d)$-design, take any two such polynomials $p(z)$ and $q(z)$. Then $f(z) = p(z) - q(z)$ is also a polynomial of degree at most $d - 1$ over $GF(n)$. Every point $(a, b)$ in the intersection $S_p \cap S_q$ satisfies $f(a) = b - b = 0$, i.e., $a$ is a root of $f$. Since no polynomial can have more roots than its degree, $|S_p \cap S_q| \leqslant d - 1$ follows.

Important combinatorial property of polynomial designs is given by the following simple proposition. We say that two points $(a_1, b_1)$ and $(a_2, b_2)$ of the grid $GF(n) \times GF(n)$ $lie$ $in$ $the$ $same$ $row$ if $a_1 = a_2$.

**Proposition 4.22** *Let $\mathcal{F}$ be a polynomial $(n, d)$-design, $S \subseteq GF(n) \times GF(n)$ be a set of $|S| \leqslant d$ points, and $\mathcal{F}_S = \{F \in \mathcal{F} : F \supseteq S\}$. Then $|\mathcal{F}_S| = n^{d-|S|}$ if no two points of $S$ lie in the same row, and $|\mathcal{F}_S| = 0$ otherwise.*

**Proof** This is a direct consequence of a standard result in polynomial interpolation. For any $l \leqslant d$ distinct points $(a_1, b_1), \ldots, (a_l, b_l)$ in $GF(n) \times GF(n)$, the number of polynomials $p(x)$ of degree at most $d - 1$ satisfying $p(a_1) = b_1, \ldots, p(a_l) = b_l$ is either 0 (if $a_i = a_j$ holds for some $i \neq j$ and hence $b_i \neq b_j$) or is exactly $n^{d-l}$:

this latter number is exactly the number of solutions of the corresponding system of linear equations, with coefficients of the polynomial $p$ viewed as variables. □

We already know that every $(\min, +)$ circuit and even every $(\min, \max, +)$ circuit approximating *minimization* problem on the polynomial $(n, d)$-design within *any* finite factor $r = r(n) \geqslant 1$ must have $n^{\Omega(d)}$ gates (Corollary 4.4). In contrast, the *maximization* problem on every polynomial $(n, d)$-design *can* be approximated within the (finite) factor $r = n/d$ by using only $\mathcal{O}(n^2)$ gates.

**Proposition 4.23** *Let $\mathcal{F}$ be a polynomial $(n, d)$-design, and $1 \leqslant d \leqslant n$. Then $\text{Max}_r(\mathcal{F}) \leqslant 3n^2$ for the factor $r = n/d$.*

**Proof** Given an input weighting $x$ of the points of the grid $GF(n) \times GF(n)$, we can first use $n(n - 1) < n^2$ max operations to compute $n$ numbers $y_1, \ldots, y_n$, where $y_i$ is the maximum weight of a point in the $i$th row of the grid. We then apply the $(\max, +)$ circuit for the top $d$-of-$n$ selection problem (see Proposition 4.11) to compute the sum $W$ of the *largest $d$* of the numbers $y_1, \ldots, y_n$ using at most $2dn \leqslant 2n^2$ additional $(\max, +)$ operations. Hence, $W$ is a sum of weights of $d$ *heaviest* points in the grid with no two points lying in the same row: each $y_i$ picks only one point in the $i$th row. Proposition 4.22 implies that these $d$ points are contained in a (unique) set of $\mathcal{F}$. Hence, the found value $W$ cannot exceed the optimal value (the weights are nonnegative). On the other hand, the weight of $d$ heaviest points of an optimal solution $F \in \mathcal{F}$ cannot exceed $W$ because $y_i$ is a *largest* weight in the entire $i$th row. Since $|F| = n$, the weight of this solution cannot exceed $(n/d) \cdot W = r \cdot W$, as desired. □

When applied to polynomial $(n, d)$-designs $\mathcal{F}$, Lemma 4.21 shows that $r = n/d$ is almost the best factor that small approximating $(\max, +)$ circuits can achieve on $\mathcal{F}$. Namely, the size of approximating $(\max, +)$ circuits drastically increases when this factor $n/d$ is only slightly decreased (halved).

**Corollary 4.24** *Let $\mathcal{F}$ be a polynomial $(n, d)$-design with $1 \leqslant d \leqslant n/2$. Then $\text{Max}_r(\mathcal{F}) \leqslant 3n^2$ for the factor $r = n/d$, but $\text{Max}_{r/2}(\mathcal{F}) \geqslant n^{d/2}$.*

In particular, by taking $d = n/2$, we have $\text{Max}_1(\mathcal{F}) \geqslant n^{n/4}$ but $\text{Max}_2(\mathcal{F}) \leqslant 3n^2$.

**Proof** The upper bound $\text{Max}_r(\mathcal{F}) \leqslant 3n^2$ for the factor $r = n/d$ is given by Proposition 4.23. On the other hand, for the factor $r/2 = n/2d$, Lemma 4.21 gives a lower bound $\text{Max}_s(\mathcal{F}) \geqslant |\mathcal{F}|/\#_l(\mathcal{F})$ with $l = d/2$. Since $|\mathcal{F}| = n^d$ and since $\#_l(\mathcal{F}) \leqslant n^{d-l}$ (by Proposition 4.22), the lower bound $\text{Max}_s(\mathcal{F}) \geqslant n^l = n^{d/2}$ follows. □

# References

1. Alon, N., Boppana, R.: The monotone circuit complexity of Boolean functions. Combinatorica **7**(1), 1–22 (1987)
2. Andreev, A.E.: On a method for obtaining lower bounds for the complexity of individual monotone functions. Sov. Math. Dokl. **31**, 530–534 (1985)

3. Crama, Y., Hammer, P.L. (eds.): Boolean functions: theory, algorithms, and applications. In: Encyclopedia of Mathematics and Its Applications, vol. 142. Cambridge University Pess, Cambridge (2011)
4. Graham, R.L., Sloane, N.J.A.: Lower bounds for constant weight codes. IEEE Trans. Inform. Theory **26**(1), 37–43 (1980)
5. Jukna, S.: Tropical complexity, Sidon sets and dynamic programming. SIAM J. Discrete Math. **30**(4), 2064–2085 (2016)
6. Jukna, S., Seiwert, H.: Approximation limitations of pure dynamic programming. SIAM J. Comput. **49**(1), 170–207 (2020)
7. Jukna, S., Seiwert, H., Sergeev, I.: Reciprocal inputs in monotone arithmetic and in tropical circuits are almost useless. Technical report, ECCC Report Nr. 170 (2020)
8. Mahajan, M., Nimbhorkar, P., Tawari, A.: Computing the maximum using (min, +) formulas. In: 42nd International Symposium on Mathematical Foundations of Computer Science (MFCS 2017), Leibniz International Proceedings in Informatics, vol. 83, pp. 74:1–74:11 (2017)
9. Newman, I., Wigderson, A.: Lower bounds on formula size of Boolean functions using hypergraph-entropy. SIAM J. Comput. **8**(4), 536–542 (1995)
10. Radhakrishnan, J.: Better lower bounds for monotone threshold formulas. J. Comput. Syst. Sci. **54**(2), 221–226 (1997)
11. Rado, R.: A theorem on independence relations. Quart. J. Math. **13**, 83–89 (1942)
12. Razborov, A.A.: Lower bounds on monotone complexity of the logical permanent. Math. Notes Acad. Sci. USSR **37**(6), 485–493 (1985)

# Chapter 5
# Tropical Branching Programs

**Abstract** Pure DP algorithms use only the basic (min, +) or (max, +) operations in their recursion equations. Many of such algorithms, especially those for various shortest path problems, are even *incremental* in that one of the inputs to the addition gate is a variable. Just as tropical circuits constitute a natural mathematical model for pure DP algorithms, tropical branching programs constitute a natural mathematical model for incremental DP algorithms. In this chapter, we prove some general lower bounds for tropical branching programs and show that for some explicit optimization problems, non-incremental pure DP algorithms *can* be super-polynomially faster than incremental algorithms. This also implies that tropical formulas can be super-polynomially weaker than tropical circuits.

## 5.1 Branching Programs

A pure DP algorithm is *incremental* if one of the two inputs to every addition (+) operation in the recursion equation is a variable. In the DP literature, incremental DP algorithms are also called *monadic*, while non-incremental ones are called *polyadic* (see, for example, the book [4]). For example, the Floyd–Warshall–Roy DP algorithm for the all-pairs shortest path problem (Example 1.8) and the DP algorithm for maximum weight independent sets in trees (Example 1.10) are not incremental. But the Bellman–Held–Karp DP algorithm (Example 1.9) for the traveling salesman problem and the Bellman–Ford–Moore DP algorithm (Example 1.7) for the single-source shortest path problem are already incremental (see Fig. 5.1). A natural mathematical model for incremental DP algorithms is that of "tropical branching programs" (tropical BPs).

A *branching program* (or a switching-and-rectifying network) $\Phi$ over a semiring $(R, \oplus, \odot)$ is a directed acyclic graph with one zero indegree node $s$ (the source node) and one zero outdegree node $t$ (the target node); programs computing several functions may have several target nodes. Multiple edges joining the same pair of nodes are allowed. Every edge is labeled by either a semiring element $c \in R$ (a "constant") or one of the variables $x_1, \ldots, x_n$. The former edges (labeled by constants) are *rectifies*, and the later edges (labeled by variables $x_i$) are *switches*.

© The Author(s), under exclusive license to Springer Nature Switzerland AG 2023
S. Jukna, *Tropical Circuit Complexity*, SpringerBriefs in Mathematics,
https://doi.org/10.1007/978-3-031-42354-3_5

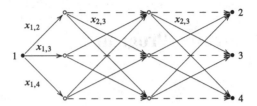

**Fig. 5.1** A tropical (min, +) branching program simulating the Bellman–Ford–Moore (min, +) incremental DP algorithm (see Example 1.7) for the shortest path problem from the source vertex $s = 1$ to all remaining vertices 2, 3, 4 of $K_4$. Dashed edges are rectifiers (labeled by constant 0). Each layer has $n - 1$ ($= 3$ in our case) nodes, and there is a switch from every node $i$ in one layer to every node $j \neq i$ in the next layer labeled by $x_{i,j}$

The polynomial *produced* by a branching program $\Phi$ is the "sum" ($\oplus$), over all $s$-$t$ paths $\pi$, of the "product" ($\odot$) of the labels of the edges along $\pi$. A branching program $\Phi$ *computes* a function $f : R^n \to R$ if the polynomial produced by $\Phi$ takes the same values as $f$ on all inputs $x \in R^n$.

As in the case of circuits, we will consider branching programs over four semirings $(R, \oplus, \odot)$: the arithmetic semiring $(\mathbb{R}_+, +, \times)$ (monotone arithmetic BPs), over two tropical semirings $(\mathbb{R}_+, \min, +)$ (tropical (min, +) BPs) and $(\mathbb{R}_+, \max, +)$ (tropical (max, +) BPs) and over the Boolean semiring $(\{0, 1\}, \vee, \wedge)$ (monotone Boolean BPs).

Note that the term produced by a path $\pi$ in a monotone *arithmetic* branching program is a term $t_\pi = c \prod_{i=1}^{n} x_i^{a_i}$, where $c$ is the product of all constant labels along $\pi$, and $a_i \in \mathbb{N}$ is the number of times the $i$th variable $x_i$ appears along $\pi$. In a *tropical* (min, +) or (max, +) branching program $\Phi$, the term produced by a path $\pi$ is a tropical term $t_\pi = a_1 x_1 + \cdots + a_n x_n + c = \langle a, x \rangle + c$, where $c$ is the sum of all constant labels along $\pi$, and each $a_i$ is (again) the number of times the $i$th variable $x_i$ appears along $\pi$. The tropical polynomial produced by the entire branching program $\Phi$ is then the minimum or the maximum of these tropical terms $t_\pi$ over all $s$-$t$ path $\pi$ in $\Phi$.

As in the case of circuits, we will concentrate not on the produced polynomials, but rather on the sets of "exponent" vectors of these polynomials. For this, we assign to each edge $e$ of $\Phi$ an *exponent vector* $z_e \in \{0, 1\}^n$ with at most one 1 by the following rule: if $e$ is a rectifier (is labeled by a constant), then $z_e := \vec{0}$ is the all-0 vectors, and if $e$ is a switch labeled by a variable $x_i$, then $z_e$ is the $i$th unit vector. The vector $b_\pi \in \mathbb{N}^n$ *produced* along a path $\pi = (e_1, e_2, \ldots, e_l)$ is then the sum $b_\pi = z_{e_1} + z_{e_2} + \cdots + z_{e_l}$ of vectors associated with its edges. Hence, the $i$th position of vector $b_\pi \in \mathbb{N}^n$ is the number of times the $i$th variable $x_i$ appears as a label of an edge of $\pi$. For example, if a path $\pi$ consists of only rectifiers, then $b_\pi = \vec{0}$ is the all-0 vector. The set $X_u \subseteq \mathbb{N}^n$ of vectors produced at a node $u \neq s$ is the set of vectors produced by the paths from the source node $s$ to $u$; for the source node $s$, we let $X_s := \{\vec{0}\}$. The set of vectors produced by the entire branching program $\Phi$ is the set $B = X_t$ of vectors produced at the target node $t$. It is easy to see that, as in the case of circuits, the set $B$ of vectors produced by a branching program $\Phi$ is the set of "exponent" vectors of the polynomial produced by $\Phi$.

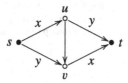

For example, the depicted branching program with 4 switches and one rectifier produces the polynomial $P = x^2 + 2xy$. The sets of vectors produced at the nodes are $X_s = \{(0, 0)\}$, $X_u = \{(1, 0)\}$, $X_v = \{(1, 0), (0, 1)\}$, and $X_t = \{(2, 0), (1, 1)\}$. Over the tropical (min, +) semiring, the branching program computes $f(x, y) = \min\{2x, x + y\}$. Note that (as in the case of circuits) the set of exponent vectors *produced* by a branching program over any semiring is always the same—it only depends on the program itself, that is, on its underlying graph and labeling of its edges, but not on the underlying semiring.

## 5.2 A Lower Bound for Thin Sets

Due to *sequential* nature of computation in branching programs (rather than *parallel* nature, as in the case of circuits), their structure is easier to analyze. For example, in the model of branching programs (instead of circuits), the following version of the lower bound given by Theorem 2.14 has a very simple proof. Recall (from Sect. 2.3) that a set $A \subseteq \mathbb{N}^n$ of vectors is $(k, l)$-*thin* for $1 \leqslant k \leqslant l$ if the following holds for any two subsets $X, Y \subseteq \mathbb{N}^n$ of vectors: if $X + Y \subseteq A$, then $|X| \leqslant k$ or $|Y| \leqslant l$.

**Lemma 5.1** *Every branching program producing a $(k, l)$-thin set $A \subseteq \mathbb{N}^n$ of size $|A| > k$ must have at least $|A|/kl$ edges.*

**Proof** Take a branching program $\Phi$ producing the set $A$, and let $E$ be the set of its edges. Define the *content* of an edge $e = (u, v)$ of $\Phi$ to be the rectangle $A_e := (X_u + \{z_e\}) + Y_v$, where $z_e \in \{0, 1\}^n$ is the exponent vector assigned to the edge $e$, $X_u \subseteq \mathbb{N}^n$ is the set of vectors produced by paths from the source node $s$ to $u$, and $Y_v \subseteq \mathbb{N}^n$ is the set of vectors produced by paths from node $v$ to the target node $t$. It is enough to show that every vector $a \in A$ belongs to the content $A_e$ of at least one edge $e \in E$ of $\Phi$ such that $|A_e| \leqslant kl$; then the desired lower bound $|E| \geqslant |A|/kl$ on the number of edges follows.

So take an arbitrary vector $a \in A$, and let $\pi$ be an $s$-$t$ path producing the vector $a$. Since $|X_s| = |\{\vec{0}\}| = 1 \leqslant k$ and $|X_t| = |A| > k$, there must be an edge $e = (u, v)$ in the path $\pi$ such that $|X_u| \leqslant k$ and $|X_v| > k$. Our vector $a$ belongs to the content $A_e = (X_u + \{z_e\}) + Y_v$ of the edge $e$. Since $X_v + Y_v \subseteq A$ and $|X_v| > k$, the $(k, l)$-thinness of $A$ implies $|Y_v| \leqslant l$. Since $|A_e| = |(X_u + \{z_e\}) + Y_v| = |X_u| \cdot |Y_v| \leqslant kl$, we are done. $\square$

**Remark 5.2** Note that the proof gives the same lower bound $|A|/kl$ also for *extended* branching programs, where instead of only 0-1 vectors $z_e$, *arbitrary* vectors $z_e \in \mathbb{N}^n$ can be assigned to the edges $e$. In this case, $|A|$ is also an *upper* bound on the number of edges in such a branching producing a set $A$: just join the vertices $s$ and $t$ by $|A|$ parallel edges, and assign each of these edges $e$ a unique vector $z_e \in A$. In particular, for $(1, 1)$-thin sets (i.e,, for Sidon sets) $A$, $|A|$ is exactly the minimum number of edges in an extended branching program producing these sets.                    □

## 5.3  Tropical Markov's Bound

The following lower bound is usually attributed to Markov [5]. A version of this bound for Boolean branching programs without rectifiers was earlier proved by Moore and Shannon [6]. The bound is also reminiscent of the min-max fact—a dual to Menger's theorem attributed to Robacker [7]: the maximum number of edge-disjoint $s$-$t$ cuts in a graph is equal to the minimum length of an $s$-$t$ path.

A *minterm* (respectively, *maxterm*) of a monotone Boolean function $f$ is a minimal set of variables such that setting these variables to 1 (respectively, to 0) forces $f$ to take value 1 (respectively, 0) regardless of the values assigned to the remaining variables.

**Lemma 5.3 (Markov [5])** *The number of switches in any monotone Boolean branching program computing a monotone Boolean function $f$ is at least $l \cdot w$, where $l$ is the minimum size of a minterm, and $w$ is the minimum size of a maxterm of $f$.*

As shown in [3], there is an analog of Markov's bound also for tropical (min, +) branching programs. Let $f(x) = \min_{a \in A} \langle a, x \rangle$ be a (min, +) polynomial with $A \subseteq \mathbb{N}^n$. Define the *width* of $f$ to be the minimum number

$$w(f) := \min\{|S| : S \subseteq [n] \text{ and } S \cap \sup(a) \neq \emptyset \text{ for all } a \in A\}$$

of positions such that every vector $a \in A$ has a nonzero entry in at least one of these positions. Note that $w(f)$ is the minimum number of variables that must be set to $+\infty$ in order to force $f$ to take value $+\infty$.

**Lemma 5.4 (Tropical Markov's Bound [3])** *Let $A \subseteq \mathbb{N}^n$ be finite with $\vec{0} \notin A$. Every (min, +) branching program computing a tropical (min, +) polynomial $f(x) = \min_{a \in A} \langle a, x \rangle$ must have at least $f(\vec{1}) \cdot w(f)$ switches.*

***Proof*** Take a (min, +) branching program $\Phi$ computing the polynomial $f$, and let $B \subseteq \mathbb{N}^n$ be the set of vectors produced by $\Phi$. Hence, $\Phi$ computes the polynomial $g(x) = \min_{b \in B} \langle b, x \rangle + c_b$ for some "coefficients" $c_b \in \mathbb{R}_+$. By Lemma 1.29, we can assume that the branching program $\Phi$ is constant-free, i.e., that no constants other than 0 are used as labels. So, $c_b = 0$ for all $b \in B$. Moreover, since $\vec{0} \notin A$, we also have $\vec{0} \notin B$.                    □

**Claim 5.5** *The program $\Phi$ has at least $g(\vec{1}) \cdot w(g)$ switches.*

**Proof** Note that the value $g(\vec{1})$ is the minimum degree $b_1 + \cdots + b_n$ of a vector $b \in B$. So, every $s$-$t$ path in the $\Phi$ must have at least $g(\vec{1}) \geqslant 1$ switches.

Define a *cut* of $\Phi$ to be a set of its switches such that every $s$-$t$ path in $\Phi$ contains at least one switch in this set. Every cut of $\Phi$ must have at least $w(g)$ switches. So, to obtain the desired lower bound $g(\vec{1}) \cdot w(g)$ on the total number of switches in $\Phi$, it is enough to show that $\Phi$ must contain at least $g(\vec{1})$ edge-disjoint cuts.

For this, associate with every node $u$ of $\Phi$ the minimum number $d_u$ of switches in a path from the source node $s$ to $u$. The source node $s$ has $d_s = 0$, and the target node $t$ has $d_t \geqslant g(\vec{1})$. Moreover, $d_v \leqslant d_u + 1$ holds for every edge $e = (u, v)$, and $d_v \leqslant d_u$ if the edge $e$ is a rectifier. For every $0 \leqslant i \leqslant d_t - 1$, let $E_i$ be the set of all edges $(u, v)$ such that $d_u = i$ and $d_v = i + 1$. Since the sets $E_i$ are clearly disjoint and all edges in $E_i$ must be switches, it remains to show that each $E_i$ is a cut. For this, take an arbitrary $s$-$t$ path $(u_1, u_2, \ldots, u_m)$ with $u_1 = s$ and $u_m = t$. The sequence of numbers $d_{u_1}, \ldots, d_{u_m}$ must reach the value $d_t \geqslant g(\vec{1})$ by starting from $d_s = 0$. At each step, the value can be increased by at most 1. So, there must be a $j$ where a jump from $d_{u_j} = i$ to $d_{u_{j+1}} = i + 1$ happens, meaning that the edge $(u_j, u_{j+1})$ belongs to $E_i$, as desired. $\square$

By Claim 5.5, it remains to show the inequalities $f(\vec{1}) \leqslant g(\vec{1})$ and $w(f) \leqslant w(g)$. Since the program $\Phi$ computes the polynomial $f$, we even have the equality $f(\vec{1}) = g(\vec{1})$. To show the inequality $w(f) \leqslant w(g)$, take any subset $S \subseteq [n]$ intersecting the supports of all vectors in $B$. It is enough to show that, then, $S$ must also intersect the supports of all vectors in $A$. Assume for the sake of contradiction that $S \cap \mathrm{sup}(a) = \emptyset$ holds for some vector $a \in A$, and take the input weighting $x \in \{0, 1\}^n$ with $x_i := 0$ for all $i \in \mathrm{sup}(a)$ and $x_i := 1$ for all $i \notin \mathrm{sup}(a)$. Then $f(x) \leqslant \langle a, x \rangle = 0$, but $g(x) \geqslant 1$ because $S \cap \mathrm{sup}(b) \neq \emptyset$ and, hence, also $\langle b, x \rangle \geqslant 1$ holds for all vectors $b \in B$, a contradiction. $\square$

Lemma 5.4 allows us to prove *tight* bounds on the number of switches in tropical branching programs solving some optimization problems.

**Corollary 5.6** *The minimum number of switches in a* (min, +) *branching program computing* $S_{n,k}(x) = \min\{\sum_{i=1}^n a_i x_i : a \in \{0, 1\}^n, a_1 + \cdots + a_n = k\}$ *is exactly* $k(n - k + 1)$.

**Proof** That $k(n - k + 1)$ switches are necessary follows from Lemma 5.4 because $S_{n,k}(\vec{1}) = k$ and $w(S_{n,k}) = n - k + 1$. That $k(n - k + 1)$ switches are also enough is shown by a branching program depicted in Fig. 5.2. $\square$

We now consider the problem of computing the sum of all entries of the product of two $n \times n$ matrices $x = (x_{i,j})$ and $y = (y_{i,j})$ over the tropical (min, +) semiring:

$$f_n(x, y) = \sum_{i, j \in [n]} \min_{k \in [n]} \{x_{i,k} + y_{k,j}\}.$$

**Corollary 5.7** *The minimum number of switches in a* (min, +) *branching program computing* $f_n$ *is exactly* $2n^3$.

***Proof*** The upper bound $2n^3$ is trivial since each minimum $g_{i,j} = \min_k\{x_{i,k} + y_{k,j}\}$ can be computed using a bunch of $2n$ switches. To prove the lower bound, we will apply Lemma 5.4.

In order to force $f_n$ to output $+\infty$, there must be at least one pair $i, j \in [n]$ such that the minimum $g_{i,j}$ outputs $+\infty$. Thus, at least $n$ variables must be set to $+\infty$, implying that $w(f_n) \geqslant n$. So, since $f_n(\vec{1}) = 2n^2$, Lemma 5.4 implies that any (min, +) branching program computing $f_n$ over the (min, +) semiring must have at least $f_n(\vec{1}) \cdot w(f_n) \geqslant 2n^3$ switches. □

## 5.4   Balanced Decompositions

Let $A \subseteq \mathbb{N}^n$ be homogeneous of degree $m$; hence, $a_1 + \cdots + a_n = m$ for all $a \in A$. A rectangle $X + Y$ is *k-homogeneous* if the set $X$ is homogeneous of degree $k$. Suppose that $A$ can be produced by a circuit (over any semiring) of size $\ell$. Then Hyafil–Valiant's decomposition of Minkowski circuits (Lemma 3.3) implies that $A$ can be written as a union of at most $\ell$ rectangles $X + Y \subseteq A$, each of which is *k-homogeneous* for *some* integer $k$ lying between $m/3$ and $2m/3$; for *different* rectangles in the decomposition of $A$, these integers $k$ can be *different*. In contrast, for branching programs, we have a much stronger result: then for *every* integer $0 \leqslant k \leqslant m$, the set $A$ can be written as a union of at most $\ell$ *k*-homogeneous (for the *same k*) rectangles.

**Lemma 5.8** *Let $A \subseteq \mathbb{N}^n$ be homogeneous of degree $m$, and suppose that $A$ can be produced by a branching program with $\ell$ nodes. Then for every $k \in [m]$, the set $A$ can be written as a union of at most $\ell$ k-homogeneous rectangles.*

***Proof*** Let $\Phi$ be a branching program (over any semiring) producing the set $A$, and let $\ell$ be the number of nodes in $\Phi$. Associate with every node $u$ of $\Phi$ the rectangle $X_u + Y_u$, where (as before) $X_u$ is the set of vectors produced by paths from the source node $s$ to $u$, and $Y_u$ is the set of vectors produced by paths from node $u$ to

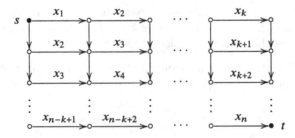

**Fig. 5.2** A branching program with $k(n - k + 1)$ switches computing the elementary symmetric polynomial $S_{n,k}(x) = \sum_{|S|=k} \prod_{i \in S} x_i$ over any semiring. The remaining edges are rectifiers labeled by the multiplicative neutral element $\mathbb{1}$ of the underlying semiring

the target node $t$. In particular, for the source node $s$ and the target node $t$, we have $X_s = \{\vec{0}\}$, $Y_s = A$, and $X_t = A$, $Y_t = \{\vec{0}\}$. Note that for every node $u$, $X_u + Y_u$ is exactly the set of all vectors produced by $s$-$t$ paths going through the node $u$. Since no vector outside $A$ can be produced, we have $X_u + Y_u \subseteq A$. So, since $A$ is homogeneous, all vectors in $X_u$ must have the *same* degree, which we denote by $d_u$, and call the *degree* of the node $u$. In particular, $d_s = 0$ and $d_t = m$ hold for the source and target gates of $\Phi$. For every edge $e = (u, v)$ of $\Phi$, we have either $d_v = d_u + 1$ (if $e$ is a switch) or $d_v = d_u$ (if $e$ is a rectifier). Since $d_s = 0$ and $d_t = m$ hold for the first and the last nodes of any $s$-$t$ path $\pi$, this means that the degrees $d_u$ of the nodes $u$ of $\pi$ take all values $0, 1, \ldots, m$. So, for every $k \in \{0, 1, \ldots, m\}$, the set $U_k = \{u : d_u = k\}$ of nodes forms a cut of the program $\Phi$, that is, every $s$-$t$ path contains at least one node $u \in U_k$. Therefore, the union of $k$-homogeneous rectangles $X_u + Y_u$ over all (at most $\ell$) nodes $u \in U_k$ gives us a covering of $A$ by at most $\ell$ $k$-homogeneous rectangles $X_u + Y_u \subseteq A$.                                                    $\square$

Lemma 5.8 only gives decompositions of sets $A \subseteq \mathbb{N}^n$ of vectors *produced* by branching programs. In general (and as in the case of tropical circuits), tropical branching programs *solving* optimization problems on sets $A$ of feasible solutions do not need to also produce these sets. But if $A$ is homogeneous and consists of only 0-1 vectors, then this is already the case (for tropical *circuits*, an analogous property was given by Lemma 1.34). A *subprogram* of a branching program is obtained by removing some of its edges.

**Lemma 5.9** *Let $A \subseteq \{0, 1\}^n$ be homogeneous. If an optimization problem on $A$ is solved by a tropical branching program $\Phi$, then some subprogram of $\Phi$ also produces the set $A$.*

**Proof** Let $A \subseteq \{0, 1\}^n$ be homogeneous of degree $m$, and consider first the *maximization* problem $f(x) = \max_{a \in A} \langle a, x \rangle$ on $A$. Let $\Phi$ be tropical (max, +) branching program solving this problem, and let $B \subseteq \mathbb{N}^n$ be the set of vectors produced by $\Phi$. Our goal is to show that some subprogram of $\Phi$ (obtained by removing some edges) must produce the set $A$.

By Lemma 1.24, we know that the inclusions $A \subseteq B \subseteq A^{\downarrow}$ hold. This implies that $A$ is the subset of all vectors in $B$ of largest degree. Apply iteratively the following transformation to the branching program $\Phi$ until possible: remove an edge $(u, v)$ if the *longest* path from the source node $s$ to node $v$ *going through* this edge is *shorter* than the longest path from $s$ to $v$. Since vectors in $A \subseteq B$ have the largest degree among all vectors in the set $B$ produced by the program $\Phi$, none of the vectors of $A$ will be lost during this removal. By applying such a removal as long as possible, we will eventually obtain the desired branching program producing the set $A$.

The argument in the case of the *minimization* problem on $A$ is almost the same. By Lemma 1.22, we then know that the inclusions $A \subseteq B \subseteq A^{\uparrow}$ hold. This implies that the set $A$ is then exactly the subset of vectors in $B$ of *smallest* degree (which is the degree of $A$). Thus, we now remove an edge $(u, v)$ if the *shortest* path from $s$ to $v$ going through this edge is *longer* than the shortest path from $s$ to $v$.        $\square$

## 5.5    Tropical BPs $\neq$ Tropical Circuits

As Lemma 5.8 shows, branching programs lead to decompositions in "strongly" balanced rectangles (with equality), while (as we have seen in Sect. 3.1) in the case of circuits, the resulting rectangles are only "weakly" balanced (with inequalities). Hrubeš and Yehudayoff [2] captured the difference between the powers of monotone arithmetic circuits and arithmetic branching programs exactly at this point.

To show such a gap, let us consider the following polynomial defined by a full binary tree. Let $n$ be of the form $n = 2^{d+1} - 1$, and let $T = (V, E)$ be a full binary (rooted) tree of depth $d$ on a set $V$ of $|V| = n$ nodes. A *labeling* of $T$ is a mapping $h : V \to \mathbb{Z}_n$ from the nodes of $T$ to the additive group $\mathbb{Z}_n = \{0, 1, \ldots, n - 1\}$ of integers modulo $n$. A labeling $h$ is *legal* if it is additive in the following sense: if $v$ is a node with children $v_1$ and $v_2$ in $T$, then $h(v) = h(v_1) + h(v_2)$. Let $H$ denote the set of all legal labelings $h : V \to \mathbb{Z}_n$ of $T$. Note that the labels of leaves in every legal labeling determine the labels of all other nodes. So, there are $|H| = n^{2^d} = n^{(n+1)/2}$ legal labelings.

The *additive tree polynomial* is the following multilinear homogeneous polynomial of degree $n$ on $n^2$ variables $x_{u,i}$ for $u \in V$ and $i \in \mathbb{Z}_n$:

$$f_n(x) = \sum_{h \in H} \prod_{u \in V} x_{u,h(u)} . \tag{5.1}$$

**Theorem 5.10 (Hrubeš and Yehudayoff [2])** *The polynomial $f_n$ can be computed by a monotone arithmetic circuit with $\mathcal{O}(n^3)$ gates, but every monotone arithmetic branching program computing $f_n$ has at least $n^{\Omega(\log n)}$ nodes.*

The two optimization problems corresponding to the polynomial $f_n$ are as follows: given an assignment of nonnegative weights $x_{u,i} \in \mathbb{R}_+$ to the points $(u, i)$ of the grid $V \times \mathbb{Z}_n$, compute the minimum or the maximum of $\sum_{u \in V} x_{u,h(u)}$ over all legal labelings $h \in H$. Since the polynomial $f_n$ is multilinear and homogeneous, Lemma 5.9 implies that the lower bound of Theorem 5.10 holds also for tropical branching programs solving the corresponding minimization and maximization problems. We thus have the same separation between tropical circuits and tropical branching programs.

**Corollary 5.11** *Both optimization problems corresponding to the polynomial $f_n$ can be solved by tropical circuits of size $\mathcal{O}(n^3)$, but every tropical branching program solving any of these two problems must have $n^{\Omega(\log n)}$ nodes.*

**Remark 5.12 (Tropical Formulas $\gg$ Tropical Circuits)** A formula is a circuit whose underlying graph is a tree; that is, gates have fanout at most 1. Since every formula is, in fact, a special branching program (when the underlying graph is parallel–sequential), Corollary 5.11 implies that tropical formulas can be super-polynomially weaker than tropical circuits.                                                    □

### 5.5.1 Proof of the Upper Bound for Circuits

Due to the "tree structure" of the monomials of $f_n$, the upper bound of Theorem 5.10 can be shown via easy dynamic programming. For $i \in \mathbb{Z}_n$, let $F^d_{r,i}$ denote the polynomial defined as $f_n$, except that the labelings $h : V \to \mathbb{Z}_n$ are now restricted to range over legal labelings with $h(r) = i$, where $r$ is the root of $T$. Hence, $f_n = \sum_{i \in \mathbb{Z}_n} F^d_{r,i}$. Observe that

$$F^d_{r,i} = x_{r,i} \cdot \sum_{j=0}^{i} F^{d-1}_{r_0,j} \cdot F^{d-1}_{r_1,i-j} ,$$

where $r_0$ and $r_1$ are the left and right children of the root $r$. Hence, if $L(d)$ denotes the number of gates required to simultaneously compute all $F^d_{r,i}$ with $i \in \mathbb{Z}_n$, then $L(d) \leqslant 2 \cdot L(d-1) + \mathcal{O}(n^2)$ with $L(0) = \mathcal{O}(n)$, which resolves to $L(d) = \mathcal{O}(n^2 2^d) = \mathcal{O}(n^3)$.

In the tropical (min, +) semiring, we have $f_n = \min_{h \in H} \sum_{u \in V} x_{u,h(u)} = \min_{i \in \mathbb{Z}_n} F^d_{r,i}$, and the recursion is $F^d_{r,i} = x_{r,i} + \min_{j+k=i} \{ F^{d-1}_{r_0,j} + F^{d-1}_{r_1,k} \}$. In the tropical (max, +) semiring, we just take maximum instead of minimum. □

### 5.5.2 Proof of the Lower Bound for Branching Programs

Let $T = (V, E)$ be a full binary tree of depth $d$ on a set $V$ of $|V| = n = 2^{d+1} - 1$ nodes and $H$ be the set of all legal labelings $h : V \to \mathbb{Z}_n$. Recall that a labeling $h : V \to \mathbb{Z}_n$ is legal if it is additive in the following sense: if $u$ is a node with children $v$ and $w$ in $T$, then $h(u) = h(v) + h(w)$. Thus, a labeling $h$ is *legal* iff the label of every node $u \in V$ is

$$h(u) = \sum_{l \in \ell(T_u)} h(l) , \qquad (5.2)$$

where $T_u$ is the subtree of $T$ rooted in $u$, and $\ell(T_u)$ is the set of leaves of this subtree. The following consequence of Eq. (5.2) will be crucial in the next proof: if two legal labelings $h \neq h'$ differ on exactly one leaf $l \in \ell(T_u)$, then $h(u) \neq h'(u)$.

Now let $\Phi$ be a monotone arithmetic branching program computing the additive tree polynomial

$$f_n(x) = \sum_{h \in H} \prod_{u \in V} x_{u,h(u)} ,$$

where $H$ is the set of all legal labelings $h : V \rightarrow \mathbb{Z}_n$ of the full binary tree $T = (V, E)$ of depth $d$. Let $s$ be the number of nodes in $\Phi$. Our goal is to show that $s \geqslant n^{\Omega(\log n)}$.

By Corollary 1.13, the program $\Phi$ also *produces* the polynomial $f_n$ (monotone arithmetic circuits and branching programs "produce what they compute"). So, since the polynomial $f_n$ is homogeneous (of degree $n$), Lemma 5.8 implies that for *every* $k \in [n]$, $f_n$ can be written as the sum $f_n = \sum_{i=1}^{s} P_i Q_i$ of $s$ or fewer products of polynomials, where each polynomial $P_i$ is homogeneous of degree $k$. In particular, if $\mathrm{mon}(f_n)$ denotes the set of all monomials of $f_n$, then $\mathrm{mon}(f_n) = \bigcup_{i=1}^{s} \mathrm{mon}(P_i Q_i)$. It is thus enough to show that for *some* $k \in [n]$, every such polynomial $P_i Q_i$ can contain at most $|H| \cdot n^{-\Omega(\log n)}$ monomials; then $s \geqslant n^{\Omega(\log n)}$ follows.

What balance parameter $k$ to choose? A nice idea of Hrubeš and Yehudayoff [2] was to look at the "isoperimetric profile" of the tree $T$ to choose such an integer $k$ accordingly. A *k-coloring* of a graph $G = (V, E)$ gives blue color to some $k$ nodes and gives red color to the remaining nodes. The *edge isoperimetric profile* of $G$ is the function $\partial_G(k)$ whose value is the minimum, over all $k$-colorings of $G$, of the number of edges whose endpoints receive different colors. The *isoperimetric peak* of the graph $G$ is $\partial_G := \max_k \partial_G(k)$ over all $1 \leqslant k \leqslant |V|$. If $T$ is a full binary tree, then $\partial_T(k)$ is small for many values of $k$. For example, if $k$ is the number of nodes in a full subtree of $T$, then $\partial_T(k) = 1$. Still, it is known that the isoperimetric peak grows with the depth of $T$. Namely, it is known that $\partial_T = \Theta(d)$ holds for the full binary tree $T$ of depth $d$ [1, 2, 8] (see also Remark 5.14).

So, fix an integer $k$ for which the isoperimetric peak of our tree $T$ is achieved; hence, $\partial_T(k) = \Omega(d) = \Omega(\log n)$. Take an arbitrary product $PQ$ of two polynomials such that $\mathrm{mon}(PQ) \subseteq \mathrm{mon}(f_n)$ holds (all monomials of $PQ$ are monomials of $f_n$), and $P$ is a homogeneous polynomial of degree $k$. Our goal is to show that, then,

$$|\mathrm{mon}(PQ)| \leqslant |\mathrm{mon}(f_n)| \cdot n^{-\partial_T(k)/4}.$$

Since the polynomial $P$ is homogeneous of degree $k$, each monomial of $P$ corresponds to a partial labeling $h : S \rightarrow \mathbb{Z}_n$ giving labels to some set $S \subseteq V$ of $k$ nodes of the tree $T$; we say that these nodes are *touched* by that monomial. Since the entire polynomial $PQ$ is multilinear (no variable has degree larger than 1), we also have that *all* monomials of the polynomial $P$ touch the *same* set $V_0 \subseteq V$ of $|V_0| = k$ nodes of our tree $T$. Hence, all monomials of $Q$ touch the set $V_1 = V \setminus V_0$ of remaining $|V_1| = n - k$ nodes of $T$.

Let $H_0$ be the set of all partial labelings $h_0 : V_0 \rightarrow \mathbb{Z}_n$ corresponding to monomials of the polynomial $P$, and $H_1$ be the set of all partial labelings $h_1 : V_1 \rightarrow \mathbb{Z}_n$ corresponding to monomials of the polynomial $Q$. Thus,

$$H_0 \circ H_1 = \{h_0 \circ h_1 : h_0 \in H_0 \text{ and } h_1 \in H_1\}$$

is the set of all (non-partial) labelings corresponding to the monomials of the polynomial $PQ$. From $\mathrm{mon}(PQ) \subseteq \mathrm{mon}(f_n)$, we have that $H_0 \circ H_1 \subseteq H$, that is, all labelings in $H_0 \circ H_1$ are legal.

Color the nodes of $V_0$ in blue, and all nodes of $V_1$ in red. Thus, blue nodes are exactly the nodes touched by the monomials of $P$, and red nodes are exactly the nodes touched by the monomials of $Q$. Let $E$ be the set of all *mixed* edges of the tree $T$, that is, of edges lying between $V_0$ and $V_1$ (endpoints of these edges have different colors). Since $|V_0| = k$, we know (by our choice of the parameter $k$) that there must be $|E| \geqslant \partial_T(k) = \Omega(\log n)$ mixed edges. So, since $|H_0 \circ H_1|$ is exactly the number of monomials in $PQ$, it remains to show the upper bound

$$|H_0 \circ H_1| \leqslant |H| \cdot n^{-|E|/4}. \tag{5.3}$$

***Proof of Eq.*** (5.3) Each legal labeling $h : V \to \mathbb{Z}_n$ of the vertices of $T$ is determined by the values it gives to the leaves of $T$. So, we will concentrate on those mixed edges $e \in E$ that are reachable by monochromatic paths from at least one leaf. Call a path from a leaf $l$ to a node $v$ *pure* if all its nodes, except the last one, have the same color and the last node $v$ has a different color. Thus, the last edge of such a path is mixed (belongs to $E$). A node $v$ is *pure* if it is the last node of a pure path from some leaf. Let $V_{\mathrm{pure}}$ be the set of all pure nodes.

For each pure node $v \in V_{\mathrm{pure}}$, fix one leaf $l_v$ of the subtree $T_v$ such that the (unique) path from $l_v$ to $v$ is pure, and call the leaf $l_v$ *pure*; there may be several leaves of $T_v$ with pure paths to the node $v$, but we pick only one of them. Note that $v$ and $l_v$ have different colors. Let $L \subseteq V$ be the set of all leaves of $T$ and $L_{\mathrm{pure}} = \{l_v \in L : v \in V_{\mathrm{pure}}\}$ be the set of all pure leaves. Since no pure path can be extended to another pure path (the last two nodes of any such path have different colors), we have $l_u \neq l_v$ for all pure nodes $u \neq v \in V_{\mathrm{pure}}$; hence, $|L_{\mathrm{pure}}| = |V_{\mathrm{pure}}|$.

The proof of Eq. (5.3) consists of two steps. First, we will show that any two different labelings from $H_0 \circ H_1$ must differ on at least one non-pure leaf $l \notin L_{\mathrm{pure}}$; this will be a direct consequence of Claim 5.13 and will give the upper bound $|H_0 \circ H_1| \leqslant |H| \cdot n^{-|L_{\mathrm{pure}}|} = |H| \cdot n^{-|V_{\mathrm{pure}}|}$. Then we will show that there must be $|V_{\mathrm{pure}}| \geqslant |E|/4$ pure nodes (as a direct consequence of Eq. (5.4)). $\qquad\square$

**Claim 5.13** *If two different legal labelings coincide on all blue nodes, then they differ on at least one red non-pure leaf.*

***Proof*** Let $h_1, h_2 : V \to \mathbb{Z}_n$ be different legal labelings such that $h_1(v) = h_2(v)$ holds for all blue nodes $v$. Since the labelings $h_1$ and $h_2$ are different but give the same color to all blue nodes (including blue leaves), they must differ on some red leaves. Suppose for the sake of contradiction that *all* these red leaves are pure, and take an arbitrary (red) pure leaf $l_v$ such that $h_1(l_v) \neq h_2(l_v)$. Since the leaf $l_v$ is red, the corresponding pure node $v$ is blue. Hence, $h_1(v) = h_2(v)$, that is, the node $v$ receives the *same* label under both labelings $h_1$ and $h_2$. Since the node $v$ is blue, every pure path that starts in a red leaf of the subtree $T_v$ must also end in a blue node of $T_v$. We can therefore assume that $v$ is *minimal* in that $h_1(l_u) = h_2(l_u)$ holds for every blue pure node $u \neq v$ of the subtree $T_v$: if not, then consider the smaller subtrees of $T_v$. Thus, we can assume that $v$ is a minimal blue node with $h_1(l_v) \neq h_2(l_v)$. Then, due to the minimality of $T_v$, the labelings $h_1$ and $h_2$ coincide on all leaves of $T_v$ with a sole exception of the (red) leaf $l_v$. But, by Eq. (5.2), this is impossible because $h_1(v) = h_2(v)$.                                                   $\square$

The number of all legal labelings of the tree is $|H| = n^{|L|}$, where (as before) $L \subseteq V$ is the set of all leaves of $T$. Let $L_0 \subseteq L$ and $L_1 \subseteq L$ be the sets of blue and red leaves. By Claim 5.13, if two distinct labelings from $H_0 \circ H_1$ coincide on all blue nodes, then they differ on at least one leaf in $L_1 \setminus L_{\text{pure}}$; hence, $|H_1| \leqslant n^{|L_1 \setminus L_{\text{pure}}|}$. By a symmetric argument, if two distinct labelings from $H_0 \circ H_1$ coincide on all red nodes, then they differ on at least one leaf in $L_0 \setminus L_{\text{pure}}$; hence, $|H_0| \leqslant n^{|L_0 \setminus L_{\text{pure}}|}$. Since $L_0 \cup L_1 = L$ and $L_0 \cap L_1 = \emptyset$, this yields the upper bound

$$|H_0 \circ H_1| \leqslant n^{|L|-|L_{\text{pure}}|} = |H| \cdot n^{-|L_{\text{pure}}|} = |H| \cdot n^{-|V_{\text{pure}}|}.$$

To finish the proof of Eq. (5.3), it therefore remains to show that there must be $|V_{\text{pure}}| \geqslant |E|/4$ pure nodes. To do this, it will be more convenient to consider not the set $V_{\text{pure}}$ of pure nodes itself, but rather the set $S$ of all nodes $v \in V$ such that the parent of $v$ in $T$ is pure. Since $|S| \leqslant 2|V_{\text{pure}}|$, it is enough to prove the lower bound

$$|S| \geqslant |E|/2. \tag{5.4}$$

To prove Eq. (5.4), let $T'$ be the minor of $T$ obtained by contracting all the edges of $T$ not in $E$. The tree $T'$ is a (not necessarily binary) tree with $|E|$ edges and, hence, $|V(T')| = |E| + 1$ nodes. View the tree $T'$ as directed from leaves to root. Let $W \subseteq V(T')$ be the set of all nodes $x \in V(T')$ such that $x$ is either a leaf of $T'$ or has indegree one in $T'$ and is not the root of $T'$. Since the tree $T'$ can have at most $|E|/2$ nodes of in-degree at least two, we have $|W| \geqslant (|E|+1)-|E|/2-1 = |E|/2$.

For a node $x \in V(T')$ of $T'$, let $[x] \subseteq V(T)$ be the set of nodes of $T$ that have been contracted to $x$, and let $v_x \in V(T)$ be the node in $[x]$ that is closest to the root of $T$. This is well defined, since the set $[x]$ is connected in $T$. So, since $v_x \neq v_y$ holds for different nodes $x \neq y$ of the tree $T'$, it is enough to show that $v_x \in S$ holds for all nodes $x \in W$; then $|S| \geqslant |W| \geqslant |E|/2$, as desired.

To show the inclusion $\{v_x : x \in W\} \subseteq S$, it is enough to show that for every node $x \in W$, there is a path from a leaf to $v_x$ in $T$, all nodes of which have the same color as $v_x$: since $x$ is not the root of $T'$, the parent of $v_x$ in $T$ is of a different color than $v_x$, meaning that $v_x \in S$. This clearly holds if $x \in W$ is a leaf of $T'$ because then all the nodes of the entire subtree $T_{v_x}$ were contracted to $x$, meaning that all these nodes have the same color as $v_x$:

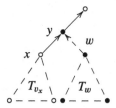

So, let $y \in W$ be a node of indegree one in $T'$. Then for some child $w$ of $v_y$ in $T$, all the nodes of the entire subtree $T_w$ were contracted to $y$, meaning that all these nodes have the same color as $v_y$. Then the parent of $v_y$ is a pure node, that is, $v_y \in S$. This completes the proof of Eq. (5.4) and, hence, also the proof of Eq. (5.3).

**Remark 5.14** Let $T$ be the full binary tree of depth $d$; hence, $T$ has $n = \sum_{i=0}^{d} 2^i = 2^{d+1} - 1$ nodes. Hrubeš and Yehudayoff [2] have not only shown that the isoperimetric peak $\partial_T :=$ $\max_k \partial_T(k)$ of $T$ is $\partial_T = \Theta(d)$ (this was already shown in [1,8]) but have also found a surprisingly tight characterization of the *entire* spectrum $\partial_T(1), \partial_T(2), \ldots, \partial_T(n)$ in terms of the following simple parameter of integers. Namely, take an integer $k \geqslant 0$, and look at its binary expansion $a_0 a_1 a_2 \ldots$ with $a_i \in \{0, 1\}$; hence, $k = \sum_{i \geqslant 0} a_i 2^i$. It is proved in [2] that for every integer $1 \leqslant k < n$, the inequalities $\frac{1}{2} \cdot \tau(k) \leqslant \partial_T(k) \leqslant 2 \cdot \tau(k)$ hold, where $\tau(k)$ denotes the number of configurations "10" (one followed by zero) in the binary expansion of $k$. In particular, $\tau(k) = 1$ holds for all integers $1 \leqslant k \leqslant n$ of the form $k = \sum_{i=0}^{m} 2^i$ for any $0 \leqslant m \leqslant d$ because their binary expansions are of the form $11 \ldots 100 \ldots$. But $\tau(k) = \Omega(d)$ for some other values of $k$. In particular, this happens for the numbers $k$ whose binary expansion is of the form $10101010 \ldots$. Say, if the depth $d = 2m$ of the tree $T$ is even, then we have $\tau(k) = m + 1 = d/2 + 1$ for $k = \sum_{i=0}^{m} 4^i = (4^{m+1} - 1)/3 = (2^{d+2} - 1)/3 = (2n + 1)/3$. If $d = 2m + 1$ is odd, then we have $\tau(k) = m + 1 = (d + 1)/2$ for $k = \sum_{i=0}^{m} 4^i = (4^{m+1} - 1)/3 = (2^{d+1} - 1)/3 = n/3$.                    □

# References

1. Bharadwaj, B.V.S., Chandran, L.S.: Bounds on isoperimetric values of trees. Discrete Math. **309**(4), 834–842 (2009)
2. Hrubeš, P., Yehudayoff, A.: On isoperimetric profiles and computational complexity. In: Proceedings of 43rd ICALP (2016), Leibniz International Proceedings in Informatics, vol. 55, pp. 89:1–89:12 (2016)
3. Jukna, S., Schnitger, G.: On the optimality of Bellman-Ford-Moore shortest path algorithm. Theor. Comput. Sci. **628**, 101–109 (2016)
4. Kumar, V., Grama, A., Karypis, G.: Introduction to Parallel Computing. Benjamin–Cumming, San Francisco (1994)
5. Markov, A.A.: Minimal relay-diode bipoles for monotonic symmetric functions. Problemy Kibernetiki **8**, 117–121 (1962). English transl. in Problems of Cybernetics **8**, 205–212 (1964)
6. Moore, E.F., Shannon, C.E.: Reliable circuits using less reliable relays. J. Franklin Inst. **262**(3), 281–297 (1956)
7. Robacker, J.T.: Min-max theorems on shortest chains and disjoint cuts of a network. Technical Report. RM-1660, The Rand Corp. (1956)
8. Vrt'o, I.: A note on isopertimetric peaks of complete trees. Discrete Math. **310**(6–7), 1272–1274 (2010)

# Chapter 6
# Extended Tropical Circuits

**Abstract** In this chapter, we first show that just one division (/) gate as the output gate can exponentially decrease the size of monotone arithmetic (+, ×) circuits. As a consequence, just one subtraction (−) gate at the "very end," as the output gate, can exponentially decrease the size of tropical (min, +) circuits. We then show that even (min, max, +) circuits (with monotone max operation instead of non-monotone subtraction) can be exponentially smaller than tropical (min, +) circuits. Finally, we show that, in contrast, subtractions at the "very beginning," at input gates, are of almost no help. Namely, the size of (min, +) circuits cannot be substantially reduced by allowing the circuits, besides input variables $x_1, \ldots, x_n$, to also use their tropical reciprocals $-x_1, \ldots, -x_n$ as inputs. The question of whether reciprocal inputs can substantially decrease the size of (min, max, +) circuits remains open.

## 6.1 Arithmetic (+, x, /) and Tropical (min, +, −) Circuits

In arithmetic (+, ×, /) circuits, besides addition (+) and multiplication(×) gates, also division (/) gates can be used. Strassen [14] has shown that if subtraction (−) gates are allowed, then division (/) gates *cannot* significantly decrease the size of arithmetic circuits: if a polynomial $f$ of degree $d$ can be computed by a (+, ×, /, −) circuit of size $s$, then $f$ can also be computed by a (+, ×, −) circuit of size $\mathcal{O}(sd \log d)$. Thus, division gates (/) *cannot* substantially decrease the size of (+, ×, −) circuits, at least of those computing polynomials of not too large degree.

But what about *monotone* arithmetic (+, ×) circuits (without subtraction gates): can then division gates (/) help? The question is not trivial because the subtraction operation (−) is crucially used in Strassen's argument to express $1/f$ as power series $1/(1-(1-f)) = \sum_{i \geq 0}(1-f)^i$. And indeed, it turned out that Strassen's result does *not* hold for monotone arithmetic (+, ×) circuits: then division (/) gates already *can* drastically decrease the size of monotone arithmetic (+, ×) circuits. This was shown by Fomin, Grigoriev, and Koshevoy [3] using directed and undirected spanning tree polynomials.

© The Author(s), under exclusive license to Springer Nature Switzerland AG 2023
S. Jukna, *Tropical Circuit Complexity*, SpringerBriefs in Mathematics,
https://doi.org/10.1007/978-3-031-42354-3_6

Let $\mathcal{T}_n$ be the family of all spanning trees $T$ of the complete graph $K_n$ on $[n] = \{1, \ldots, n\}$; recall that a *spanning tree* of $K_n$ is a tree on $[n]$ which includes all $n$ vertices of $K_n$. An *arborescence* (or a *directed spanning tree*) on $[n]$ is a directed tree $T$ on $[n]$ such that the vertex $r = 1$ (the root) is reachable from every other vertex. Let $\vec{\mathcal{T}}_n$ be the family of all arborescences on $[n]$.

Associate with each (directed or undirected) edge $e$ between two vertices in $[n]$ a (formal) variable $x_e$, interpreted as the weight of this edge. The *spanning tree polynomial* $\mathrm{ST}_n$ and the *directed spanning tree polynomial* $\mathrm{DST}_n$ on $[n]$ are the following homogeneous multilinear polynomials of degree $n - 1$:

$$\mathrm{ST}_n(x) = \sum_{T \in \mathcal{T}_n} \prod_{e \in T} x_e \quad \text{and} \quad \mathrm{DST}_n(x) = \sum_{T \in \vec{\mathcal{T}}_n} \prod_{e \in T} x_e. \tag{6.1}$$

For definiteness, if $n = 1$ (just a single vertex), then $\mathrm{ST}_n(x)$ and $\mathrm{DST}_n(x)$ are defined to be the constant 1 polynomial. We already known that $\mathrm{ST}_n$ requires monotone arithmetic $(+, \times)$ circuits of size $2^{\Omega(\sqrt{n})}$ (Theorem 3.16) and that $\mathrm{DST}_n$ requires monotone arithmetic $(+, \times)$ circuits of size $2^{\Omega(n)}$ (Theorem 3.13). In contrast, monotone arithmetic $(+, \times, /)$ circuits (with division gates) for these polynomials are even exponentially smaller.

**Theorem 6.1 (Fomin, Grigoriev, and Koshevoy [3])** *Both polynomials* $\mathrm{ST}_n$ *and* $\mathrm{DST}_n$ *can be computed by arithmetic* $(+, \times, /)$ *circuits of size* $\mathcal{O}(n^3)$.

The authors of [3] have first shown that a slightly weaker upper bound $\mathcal{O}(n^4)$ for $\mathrm{ST}_n$ (undirected case) follows fairly easily from two classical results established by electrical engineers at least 100 years ago: the *Kirchhoff's effective conductance formula* [7] and the machinery of *star-mesh transformation* (see, for example, [2, Corolary 4.21]). A proof of a better upper bound $\mathcal{O}(n^3)$ was presented in [3] only for the directed spanning tree polynomial $\mathrm{DST}_n$. But (as mentioned in [3]) the same argument works also in the undirected case by viewing each undirected edge $\{i, j\}$ as a pair of directed edges $(i, j)$ and $(j, i)$ of the same weight. Bellow, we will follow an adoption to undirected case suggested by Igor Sergeev (private communication).

As in [3], we will use the well-known tree enumerator polynomial. Recall that the degree of a vertex in a tree (or in any graph) is the number of edges containing this vertex. We will determine the generating function enumerating labeled trees on the vertex set $[n] = \{1, 2, \ldots, n\}$, weighted by their vertex degrees. Thus, we introduce variables $x_1, \ldots, x_n$ corresponding to the vertices, and associate with every tree $T$ the monomial $x_T = x_1^{d_1} x_2^{d_2} \cdots x_n^{d_n}$, where $d_i$ is the degree of vertex $i$ in $T$. Note that we can also write the same monomial as the product $x_T = \prod_{\{i,j\} \in T} x_i x_j$ over all edges of $T$. This is the same thing because each variable $x_i$ appears once in the above product for every edge of $T$ that contains the vertex $i$. The *tree enumerator* is the polynomial

$$C_n(x_1, \ldots, x_n) = \sum_{T \in \mathcal{T}_n} x_T = \sum_{T \in \mathcal{T}_n} \prod_{\{i,j\} \in T} x_i x_j.$$

Note that $C_n(x) = \mathrm{ST}_n(z)$ is the spanning tree polynomial of $K_n$ under the special (product) weighting $z_{i,j} := x_i x_j$ of the edges $\{i, j\}$ of $K_n$ determined by the weights

$x_i$ and $x_j$ given to their endpoints. Also note that $C_n(\vec{1})$ is the number $|\mathcal{T}_n|$ of all spanning trees in $K_n$. As a byproduct, the following lemma yields $|\mathcal{T}_n| = n^{n-2}$.

**Lemma 6.2 (Cayley [1])** $C_n(x_1, \ldots, x_n) = x_1 x_2 \cdots x_n (x_1 + x_2 + \cdots + x_n)^{n-2}$.

*Proof* The complete graph $K_n$ of $[n] = \{1, \ldots, n\}$ is labeled, that is, each vertex has its unique number. As shown by Prüfer [12], every spanning tree $T$ of $K_n$ is uniquely specified by the sequence $a_1, a_2, \ldots, a_{n-2}$ of (not necessarily distinct) numbers $a_i \in [n]$, called the *Prüfer code* of $T$. Let $a_1$ be the number of a vertex which is adjacent to a leaf of $T$ with the *smallest* number. Remove that leaf together with the edge joining it with vertex $a_1$, do the same to determine the second number $a_2$, and repeat until one edge remains. For example, the Prüfer code of a spanning tree of $K_7$ depicted below is $(2, 7, 5, 5, 2)$ and the last edge is $\{2, 7\}$. It is not difficult to show (say, by induction on $n$) that every vertex $i \in [n]$ appears in the Prüfer code of $T$ exactly $d_i - 1$ times, where $d_i$ is the degree of $i$ in $T$. Thus, a spanning tree with a Prüfer code $a_1, a_2, \ldots, a_{n-2}$ appears in the Cayley polynomial $C_n(x)$ as the monomial $x_1 x_2 \cdots x_n \prod_{i=1}^{n-2} x_{a_i}$.

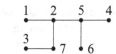

□

***Proof of Theorem 6.1*** We construct the desired circuit for the spanning tree polynomial $\mathrm{ST}_{K_n}(x) := \mathrm{ST}_n(x)$ of the compete graph $K_n$ on $[n] = \{1, \ldots, n\}$ inductively. Consider a new weighted graph $K'_{n-1}$ on the vertex set $[n-1] = \{1, \ldots, n-1\}$ obtained from the graph $K_n$ as follows: remove the vertex $n$ from $K_n$ together with all its incident edges, then draw a new (parallel) edge $e_{i,j}$ between every pair $i \neq j$ of vertices in $[n-1]$, and give this edge the weight

$$x'_{i,j} := \frac{x_{i,n} \cdot x_{j,n}}{X_n}, \tag{6.2}$$

where

$$X_n := \sum_{i \in [n-1]} x_{i,n}$$

is the sum of weights of edges incident to the vertex $n$ in $K_n$. Thus, in the resulting weighted graph $K'_{n-1}$, every two vertices $i \neq j \in [n-1]$ are joined by two parallel edges, an *old* edge $\{i, j\}$ of $K_n$ with weight $x_{i,j}$, and a *new* edge $e_{i,j}$ with weight $x'_{i,j}$. Our goal is to prove the following equality:

$$\mathrm{ST}_{K_n}(x) = X_n \cdot \mathrm{ST}_{K'_{n-1}}(x, x'). \tag{6.3}$$

□

We will first finish the proof of Theorem 6.1 using Eq. (6.3) and then will prove
Eq. (6.3) itself.

**Proof of Theorem 6.1 given Eq.** (6.3)  In the graph $K'_{n-1}$, we have two parallel
edges between each two vertices $i \neq j$ of $K_{n-1}$, one (old) with the (old) weight $x_{i,j}$
and the other (new) edge with the (new) weight $x'_{i,j} = x_{i,n}x_{j,n}/X_n$. We can replace
these two parallel edges by one edge with the accumulated weight $y_{i,j} := x_{i,j} + x'_{i,j}$:
the spanning tree polynomial $\text{ST}_{K'_{n-1}}(x, x')$ on $K'_{n-1}$ remains the same. Thus, we
actually have the equality $\text{ST}_{K_n}(x) = X_n \cdot \text{ST}_{K_{n-1}}(y)$, where $\text{ST}_{K_{n-1}}(y)$ is the
spanning tree polynomial of the weighted graph $K_{n-1}$ with new edge weights
$y_{i,j} = x_{i,j} + x_{i,n}x_{j,n}/X_n$ for all $i \neq j \in [n-1]$. Now, given a $(+, \times, /)$ circuit
$\Phi(y)$ for the spanning tree polynomial $\text{ST}_{K_{n-1}}(y)$ of $K_{n-1}$, we can compute the
spanning tree polynomial $\text{ST}_{K_n}(x)$ of $K_n$ using additional $\mathcal{O}(n^2)$ gates to compute
all weights $y_{i,j} := x_{i,j} + x_{i,n}x_{j,n}/X_n$ which are then given as inputs of the circuit
$\Phi(y)$. Thus, since we have only $n$ vertices in $K_n$, the polynomial $\text{ST}_{K_n}(x)$ can be
computed using $n \cdot \mathcal{O}(n^2) = \mathcal{O}(n^3)$ gates.                                $\square$

**Proof of Eq.** (6.3)  We will write both polynomials $\text{ST}_{K_n}$ and $\text{ST}_{K'_{n-1}}$ as sums of
polynomials, the monomials of which correspond to spanning trees associated with
different partitions of the vertex set $[n-1]$. Note that each spanning tree $T$ of $K_n$
consists of some number $m$ ($1 \leqslant m \leqslant n-1$) of *vertex-disjoint* subtrees $T_1, \ldots, T_m$
of $T$ joined by $m$ edges to the vertex $n$ in the tree $T$.

So, let $\mathcal{A} = \{A_1, \ldots, A_m\}$ be a partition of $[n-1]$ for some $1 \leqslant m \leqslant n-1$.
We associate a spanning tree $T$ of $K_n$ with this partition if the degree of the vertex
$n$ in the tree $T$ is $m$ (the number of blocks in the partition $\mathcal{A}$), and $A_1, \ldots, A_m$ are
sets of vertices of vertex-disjoint subtrees $T_1, \ldots, T_m$ of $T$, where each subtree $T_k$
is connected in $T$ to the vertex $n$ by some edge $\{i, n\}$ with $i \in A_k$ of weight $x_{i,n}$ (see
Fig. 6.1). Thus, if $K_A$ denotes the complete graph on $A \subseteq [n]$, then

$$\text{ST}_{K_n}(x) = \sum_{\mathcal{A}} \prod_{A \in \mathcal{A}} X_A \cdot \text{ST}_{K_A}(x) = \sum_{\mathcal{A}} \prod_{A \in \mathcal{A}} X_A \prod_{A \in \mathcal{A}} \text{ST}_{K_A}(x), \qquad (6.4)$$

where the sum is over all partitions $\mathcal{A}$ of the vertex set $[n-1]$, and

$$X_A := \sum_{i \in A} x_{i,n} \, .$$

In particular, if $|\mathcal{A}| = 1$ (that is, if the partition $\mathcal{A}$ consists of just one block $[n-1]$),
then the corresponding to $\mathcal{A}$ terms are those of the polynomial $X_n \cdot \text{ST}_{K_{n-1}}(x)$.

Let us now look at the form of the polynomial $\text{ST}_{K'_{n-1}}$. We associate a spanning
tree $T$ of $K'_{n-1}$ with a partition $\mathcal{A} = \{A_1, \ldots, A_m\}$ of $[n-1]$ if $A_1, \ldots, A_m$ are
the sets of vertices of subtrees $T_1, \ldots, T_m$ obtained from the tree $T$ after all its new
edges $e_{i,j}$ are removed from $T$ (see Fig. 6.1). In the tree $T$, the subtrees $T_1, \ldots, T_m$
of $T$ are connected by the edges of some spanning tree of the complete graph $K_{\mathcal{A}}$
(the *block graph* on the partition $\mathcal{A}$) whose vertices are the blocks $A_1, \ldots, A_m$ of

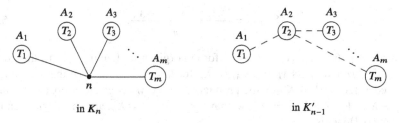

**Fig. 6.1** Structure of spanning trees in $K_n$ and in $K'_{n-1}$ associated with a given partition $\mathcal{A} = \{A_1, \ldots, A_m\}$ of $[n-1] = \{1, \ldots, n-1\}$, where $A_i$ is the set of vertices of the subtree $T_i$. Dashed lines are edges of a spanning tree of the complete "block graph" $K_{\mathcal{A}}$ whose vertices are the blocks $A_i$ of the partition $\mathcal{A}$. These dashed edges are from the set of new edges $e_{i,j}$ (with new weights $x'_{i,j}$) added when going from the graph $K_n$ on $[n]$ to the graph $K'_{n-1}$ on $[n-1]$

the partition $\mathcal{A}$. Each new edge $e_{i,j}$ of $K'_{n-1}$ has weight $x'_{i,j}$, as given by Eq. (6.2). We give each edge $\{A_k, A_l\}$ of the block graph $K_{\mathcal{A}}$ (if there is any, that is, if $m \geqslant 2$) the weight

$$z_{k,l} := \sum_{i \in A_k, j \in A_l} x'_{i,j} = \frac{X_{A_k} \cdot X_{A_l}}{X_n} = \frac{1}{X_n} \left( \sum_{i \in A_k} x_{i,n} \right) \left( \sum_{j \in A_l} x_{j,n} \right). \quad (6.5)$$

Thus, the spanning tree polynomial of the block graph $K_{\mathcal{A}}$ with $|\mathcal{A}| \geqslant 2$ vertices is of the form

$$ST_{K_{\mathcal{A}}}(x') = \sum_T \prod_{\{A_k, A_l\}} z_{k,l} = \frac{1}{X_n} \sum_T \prod_{\{A_k, A_l\}} \left( \sum_{i \in A_k} x_{i,n} \right) \left( \sum_{j \in A_l} x_{j,n} \right),$$

where the sums are over all spanning trees $T$ the block graph $K_{\mathcal{A}}$, and the product is over all edges $\{A_k, A_l\}$ of $T$. If $|\mathcal{A}| = 1$, that is, if the partition $\mathcal{A}$ consists of just one block $[n-1]$ (the graph $K_{\mathcal{A}}$ has no edges at all), then $ST_{K_{\mathcal{A}}} = 1$. The spanning tree polynomial of the entire weighted graph $K'_{n-1}$ is then of the form

$$ST_{K'_{n-1}}(x, x') = \sum_{\mathcal{A}} ST_{K_{\mathcal{A}}}(x') \prod_{A \in \mathcal{A}} ST_{K_A}(x), \quad (6.6)$$

where, as in Eq. (6.4), the sum is over all partitions $\mathcal{A}$ of the vertex set $[n-1]$. That is, for every partition $\mathcal{A} = \{A_1, \ldots, A_m\}$ of $[n-1]$, we first pick some spanning trees $T_1, \ldots, T_m$ on $A_1, \ldots, A_m$, then pick a spanning tree of the block graph $K_{\mathcal{A}}$, and for each its edge $\{A_k, A_l\}$, pick an edge $e_{i,j} \in A_k \times A_l$ of the graph $K'_{n-1}$ of weight $x'_{i,j} = \frac{x_{i,n} \cdot x_{j,n}}{X_n}$ joining some vertex $i \in A_k$ with some vertex $j \in A_l$.

By comparing the polynomials in Eqs. (6.4) and (6.6), we see that the desired equality Eq. (6.3) holds if for every partition $\mathcal{A} = \{A_1, \ldots, A_m\}$ of $[n-1]$, the spanning tree polynomial $ST_{K_{\mathcal{A}}}(x')$ of the block graph $K_{\mathcal{A}}$ satisfies

$$ST_{K_A} = \frac{X_{A_1} \cdot X_{A_2} \cdots X_{A_m}}{X_n}. \tag{6.7}$$

To verify Eq. (6.7), apply the Cayley formula (given by Lemma 6.2) to the graph $K_A$ with *vertex* weights $y_k := X_{A_k}/\sqrt{X_n}$ for $k = 1, \ldots, m$. Note that the weight of an edge $\{A_k, A_l\}$ of $K_A$ is the product $z_{k,l} = y_k y_l$, as given in Eq. (6.5). Since $X_{A_1} + X_{A_2} + \cdots + X_{A_m} = X_n$, we have $\sum_{k=1}^{m} y_i = \sqrt{X_n}$, and the Cayley lemma (Lemma 6.2) yields

$$ST_{K_A} = \prod_{k=1}^{m} y_i \left( \sum_{k=1}^{m} y_i \right)^{m-2} = \frac{X_{A_1} \cdots X_{A_m}}{(X_n)^{m/2}} \cdot (X_n)^{m/2-1} = \frac{X_{A_1} \cdots X_{A_m}}{X_n},$$

as desired. This completes the proof of Eq. (6.3) and, hence, also the proof of Theorem 6.1. □

**Remark 6.3 (One Division Gate is Enough)** The upper bound $\mathcal{O}(n^3)$ for the spanning tree polynomial $ST_G$ shown in Theorem 6.1 also holds when $(+, \times, /)$ circuits are allowed to use only one division gate at the end: one can easily move all division gates to the output gate via $(x/y) + z = (x + yz)/y$, $(x/y)z = (xz)/y$ and $(x/y)/z = x/(yz)$. Thus, the $(+, \times)/(+, \times, /)$ gap *can* be exponential even when only a single division operation "at the very end" (at the output gate) is allowed. □

Recall that the MST problem (the minimum weight spanning tree problem) on a graph $G$ is to compute $\tau(x) = \min_T x(T)$ for all input weightings $x : E \to \mathbb{R}_+$, where $x(T) = \sum_{e \in T} x(e)$, and the minimum is over all spanning trees $T$ of $G$. A spanning tree $T$ of $G$ is *x-optimal* if $x(T) = \tau(x)$.

We know (see Theorem 3.16) that any $(\min, +)$ circuit solving the MST problem on $K_n$ must have at least $2^{\Omega(\sqrt{n})}$ gates. On the other hand, Theorem 6.1 implies that $(\min, +, -)$ circuits (with subtraction gates) can easily solve this problem.

**Corollary 6.4** *The MST problem on any connected subgraph $G$ of $K_n$ can be solved by a $(\min, +, -)$ circuit of size $\mathcal{O}(n^3)$.*

**Proof** The MST problem on the entire graph $K_n$ is the tropical $(\min, +)$ version of the spanning tree polynomial $ST_n$. By Theorem 6.1, the polynomial $ST_n$ can be computed by an arithmetic $(+, \times, /)$ circuit $\Phi$ of size $\mathcal{O}(n^3)$ (this circuit has no constants as inputs.) Since arithmetic division $x/y$ in tropical $(\min, +)$ and $(\max, +)$ semirings turns[1] into (also arithmetic) subtraction $x - y$, the arithmetic $(+, \times, /)$ circuit $\Phi$ turns into a tropical $(\min, +, -)$ circuit $\Phi'$ of size $\mathcal{O}(n^3)$ solving the MST problem on $K_n$.

The tropical circuit $\Phi'$ can be used to solve the MST problem on *any* connected subgraph $G = (V, E)$ of $K_n$ (not only on the entire graph $K_n$) using additional $\mathcal{O}(n^2)$ gates. Namely, for an input weighting $x : E \to \mathbb{R}_+$, compute the sum $M = \sum_{e \in E} x(e)$ using $|E| = \mathcal{O}(n^2)$ addition gates, and give the weight $x(e) := M + 1$ to every non-edge $e \notin E$ of $G$. Under the resulting weighting, every minimum weight spanning tree of $K_n$ is a minimum weight spanning tree of the graph $G$ under the original weighting. □

## 6.2 Tropical (min, max, +) Circuits

Corollary 6.4 shows that subtraction $(-)$ gates can exponentially speed up tropical
(min, +) circuits and, hence, also pure dynamic programs. It turns out that already
a monotone max operation instead of subtraction (a non-monotone operation) can
do this: the MST problem on every $n$-vertex graph $G$ can also be solved by
a (min, max, +) circuit of size only $\mathcal{O}(n^3)$. Note that (min, +, −) circuits can
simulate (min, max, +) circuits via $\max\{x, y\} = -\min\{-x, -y\}$, but not vice
versa.

**Theorem 6.5 (Jukna and Seiwert [5])** *The MST problem on every undirected
connected $n$-vertex graph $G = (V, E)$ can be solved by a* (min, max, +) *circuit
of size* $\mathcal{O}(n^3)$.

The idea is to relate the MST problem to the min-max distances between vertices.
Given an assignment $x : E \to \mathbb{R}_+$ of nonnegative weights $x(e)$ to the edges of $G$, the
*min-max distance*, denoted as $x[u, v]$, between two vertices $u$ and $v$ is the minimum,
over all paths from $u$ to $v$ in $G$, of the maximum weight of an edge along this path.
Define the *min-max weight* $x[e]$ of an edge $e = \{u, v\}$ to be the min-max distance
$x[u, v]$ between its endpoints $u$ and $v$. That is,

$$x[e] := \min_{P} \max_{f \in P} x(f),$$

where the minimum is taken over all paths $P \subseteq E$ in $G$ between the endpoints $u$
and $v$ of the edge $e$. Note that $x[e] \leqslant x(e)$ holds for every edge $e \in E$: the edge $e$
itself is also a path between its endpoints.

**Remark 6.6** Min-max weights $x[e]$ of *all* edges $e \in E$ can be simultaneously
computed by (min, max) circuit of size $\mathcal{O}(n^3)$ resulting from the Floyd–Warshall–
Roy DP algorithm (see Example 1.8).                                              □

The following lemma relates the min-max weights of edges with the edge-
exchange property of spanning trees. Given a spanning tree $T$ of a graph $G = (V, E)$, we say that an edge $f \in T$ *can be replaced* by an edge $e \in E$ or,
equivalently, that the edge $e$ *can replace* the edge $f$ in $T$ if the graph $T - f + e$
(obtained by removing the edge $f$ from $T$ and adding the edge $e$) is also a spanning
tree of $G$.

---

[1] If elements $x \in R$ of a semiring $(R, \oplus, \odot)$ with multiplicative neutral element $\mathbb{1} \in R$ have their
inverses (or reciprocals) $x^{-1} \in R$ satisfying $x \odot x^{-1} = x^{-1} \odot x = \mathbb{1}$, then the "division" operation
$x \oslash y$ is defined by $x \oslash y := x \odot y^{-1}$. In tropical semirings $(\mathbb{R}_+, \min, +)$ and $(\mathbb{R}_+, \max, +)$,
"multiplicative" neutral element $\mathbb{1}$ is 0 (because $x + 0 = 0 + x = x$). This means that arithmetic
reciprocals $x^{-1} = 1/x$ turn into tropical reciprocals $0 - x = -x$. Thus, arithmetic division $x/y$
turns into (also arithmetic) subtraction $x - y$.

**Lemma 6.7 (Maggs and Plotkin [9])** *Let $G = (V, E)$ be an undirected connected graph, $x : E \to \mathbb{R}_+$ be a weighting, and $T \subseteq E$ be a spanning tree of $G$.*

 (i) *Every edge $e \in E$ can replace some edge $f \in T$ with $x(f) \geqslant x[e]$.*
 (ii) *Every edge $e \in T$ can be replaced by some edge $f \in E$ with $x(f) \leqslant x[e]$.*
 (iii) *If $T$ is $x$-optimal, then $x[e] = x(e)$ holds for all edges $e \in T$. If, furthermore, all weights are distinct, then also $x[e] < x(e)$ holds for all edges $e \notin T$.*

***Proof*** To show claim (i), take an edge $e \in E$ of $G$. If $e \in T$, then the edge $e$ can be replaced by itself. Now suppose that $e \notin T$, and let $P$ be the (unique) path in $T$ between the endpoints of $e$. Let $f \in P$ be an edge along that path of maximal weight $x(f)$. By the definition of $x[e]$, *every* path in $G$ between the endpoints of $e$ (including the path $P$) must contain an edge of $x$-weight at least $x[e]$. Hence, $x(f) \geqslant x[e]$. The removal of the edge $f$ from $T$ cuts the tree $T$ into two connected components. Since the set $P + e$ forms a cycle, the edge $e$ lies between these two components. Thus, $T - f + e$ is a spanning tree of $G$.

To show claim (ii), take an edge $e \in T$, and let $P$ be a path in $G$ between the endpoints of $e$ on which $x[e] = \max_{f \in P} x(f)$ is achieved. In particular, $x(f) \leqslant x[e]$ holds for all edges $f \in P$. The path $P$ does not need to lie in the tree $T$, but at least one edge $f \in P$ must cross the cut induced by the edge $e$ of $T$, that is, must lie between the two connected components of $T$ after the edge $e$ is removed. Thus, $T - e + f$ is also a spanning tree of $G$.

To show claim (iii), let $T$ be an $x$-optimal spanning tree of $G$, and take any of its edges $e \in T$. If $x[e] < x(e)$, then take an edge $f \in E$ guaranteed by Lemma 6.7(ii) such that $x(f) \leqslant x[e] < x(e)$ and $T' = T - e + f$ is also a spanning tree of $G$. But $x(T') = x(T) - x(e) + x(f) < x(T)$, contradicting the optimality of the tree $T$. Thus, $x[e] = x(e)$ holds for all edges $e \in T$.

Now suppose that all weights $x(e)$ are distinct, but $x[e] = x(e)$ holds for some edge $e \notin T$. Take an edge $f \in T$ guaranteed by Lemma 6.7(i) such that $x(f) \geqslant x[e] = x(e)$ and $T' = T - f + e$ is also a spanning tree of $G$. Since $x(f) \neq x(e)$, we have $x(f) > x(e)$ and, hence, also $x(T') = x(T) - x(f) + x(e) < x(T)$, contradicting the optimality of the tree $T$.                                     □

**Remark 6.8** If a weighting $x : E \to \mathbb{R}_+$ gives different weights to different edges, then the $x$-optimal spanning tree $T_x$ is unique and, by Lemma 6.7(iii), is of the form $T_x = \{e \in E : x[e] = x(e)\}$. Hence, in this case, we can use Floyd–Warshall–Roy (min, max) circuit of size $\mathcal{O}(n^3)$ to compute min-weights $x[e]$ of all edges $e \in E$, remove all edges $e \in E$ with $x[e] < x(e)$, and add up the weights of the remaining edges. But this requires *conditional branchings*: if $x[e] = x(e)$, then accept $e$, else reject $e$. Thus, the resulting (min, max, +) DP algorithm (with branchings) cannot be turned into a (min, max, +) circuit. Lemma 6.10 below removes the need of conditional branchings (also, which is already less important, the weights need not be distinct).        □

Let $G = (V, E)$ be an undirected connected graph and $x : E \to \mathbb{R}_+$ be a weighting of its edges. If we drop down to 0 the weight $x(e)$ of a single edge $e$ of $G$, by how much then the minimum weight of a spanning tree of $G$ drops down? The following lemma shows that the drop-down is by exactly the min-max weight $x[e]$ of this edge.

**Lemma 6.9** *Let $x : E \to \mathbb{R}_+$ be a weighting, $e \in E$ be an edge, and $x' : E \to \mathbb{R}_+$ be the weighting obtained from $x$ by giving zero weight to the edge $e$, and leaving other weights unchanged. Then $\tau(x) - \tau(x') = x[e]$.*

**Proof** It is enough to show the inequalities $\tau(x') \leqslant \tau(x) - x[e]$ and $\tau(x) \leqslant \tau(x') + x[e]$. To show the former inequality, let $T$ be an $x$-optimal spanning tree of $G$. If $e \in T$, then $x'(T) = x(T) - x(e)$. Since $x[e] \leqslant x(e)$ and $x(T) = \tau(x)$, inequality $\tau(x') \leqslant x'(T) = x(T) - x(e) \leqslant \tau(x) - x[e]$ trivially holds in this case. If $e \notin T$, then by Lemma 6.7(i), there is an edge $f \in T$ such that $x(f) \geqslant x[e]$ and $T^* = T - f + e$ is a spanning tree of $G$. Thus, also in this case (when $e \notin T$), the desired inequality follows: $\tau(x') \leqslant x'(T^*) = x'(T) - x'(f) + x'(e) = x(T) - x(f) \leqslant x(T) - x[e] = \tau(x) - x[e]$.

To show the second inequality $\tau(x) \leqslant \tau(x') + x[e]$, we use the fact that the $x'$-weight $x'(e) = 0$ of the edge $e$ is the *smallest* possible weight (all weights are nonnegative). So, by Lemma 6.7(i) applied to the weighting $x'$, we have that $e \in T$ holds for at least one $x'$-optimal spanning tree $T$ of $G$. By Lemma 6.7(ii) applied to the tree $T$ under the weighting $x$, there is an edge $f$ of $G$ of weight $x(f) \leqslant x[e]$ such that $T^* = T - e + f$ is a spanning tree of $G$. So, since $x'(T) = x(T - e)$ holds, the desired inequality $\tau(x) \leqslant x(T^*) = x(T - e) + x(f) = x'(T) + x(f) \leqslant x'(T) + x[e] = \tau(x') + x[e]$ follows. □

We will use the following direct consequence of Lemma 6.9.

**Lemma 6.10** *Let $G = (V, E)$ be an undirected connected $n$-vertex graph and $T = \{e_1, \ldots, e_r\}$ be a spanning tree of $G$; hence, $r = n - 1$. Then for every weighting $x : E \to \mathbb{R}_+$, we have $\tau(x) = x_0[e_1] + x_1[e_2] + \cdots + x_{r-1}[e_r]$, where $x_0 = x$, and each next weighting $x_i : E \to \mathbb{R}_+$ is obtained from $x$ by setting the weights of edges $e_1, \ldots, e_i$ to zero.*

**Proof** Lemma 6.9 gives us the recursion $\tau(x_i) = \tau(x_{i+1}) + x_i[e_{i+1}]$, which rolls out into $\tau(x) = \tau(x_r) + x_{r-1}[e_r] + \cdots + x_1[e_2] + x_0[e_1]$. Since the weighting $x_r$ gives weight 0 to all edges $e_1, \ldots, e_r$ of the tree $T$, we have $\tau(x_r) = x_r(T) = 0$; recall that all weights are *nonnegative*. □

**Proof of Theorem 6.5** It is enough to show that the MST problem on the complete graph $K_n$ can be solved by a (min, max, +) circuit of size $\mathcal{O}(n^3)$: as shown in the proof of Corollary 6.4, this circuit can be easily turned into a (min, max, +) circuit of size $\mathcal{O}(n^3)$ solving the MST problem on any given subgraph $G$ of $K_n$.

According to Lemma 6.10, we only have to compute min-max weights of $n - 1$ edges $e_1, \ldots, e_{n-1}$ (of a fixed in advance spanning tree $T$) under the $n - 1$ weightings $x_0 = x, x_1, \ldots, x_{n-2}$ and add these min-max weights together. For the $i$th weighting $x_i$ (obtained from the initial input weighting $x \in \mathbb{R}_+^n$ by setting the weights of $e_1, \ldots, e_{i-1}$ to 0), the min-max weight $x_i[e_{i+1}]$ of the edge $e_{i+1}$ under the weighting $x_i$ can be computed by a (min, max) circuit of size $\mathcal{O}(n^3)$ resulting from the Bellman–Ford–Moore DP algorithm (see Example 1.7). This gives us a (min, max, +) circuit of size $n \cdot \mathcal{O}(n^3) = \mathcal{O}(n^4)$ solving the MST problem on any $n$-vertex graph. But since in our case each next weighting $x_i$ differs from the previous weighting $x_{i-1}$ on only one edge, the total number of operations can be reduced till $\mathcal{O}(n^3)$.

First, compute all min-max distances $x[i, j]$ under the initial weighting $x$ using the Floyd–Warshall–Roy (min, max) circuit of size $N = \mathcal{O}(n^3)$ (see 1.8). After that, it is enough just to update these weights. Namely, the next to $x$ weighting $x'$ only sets the weight of one edge $e = \{a, b\}$ (of the fixed spanning tree $T$) to 0 and leaves the weights of other edges unchanged. We thus only have to update the max-lengths of paths going through the edge $e$. A path between vertices $i$ and $j$ justifying the min-max distance $x'[i, j]$ either goes from the edge $e = \{a, b\}$ or not. So, $x'[i, j]$ is the minimum of $x[i, j]$ and of $\max\{x[i, a], x[b, j]\}$ and $\max\{x[i, b], x[a, j]\}$. We thus can compute the min-max distances between all pairs of vertices under the next to $x$ weighting $x'$ using only $K = O(n^2)$ additional (min, max) operations. Since, by Lemma 6.10, we only need to compute min-max distances of $n - 1$ edges with respect to $n - 1$ weightings $x_0, \cdots x_{n-2}$ obtained from a given weighting, the total number of performed operations is at most $N + (n - 2)K = O(n^3)$.                    □

## 6.3  Reciprocal Inputs Cannot Help Much

Corollary 6.4 shows that the size of tropical (min, +) circuits can be exponentially decreased by allowing just one subtraction "at the very end," at the output gate. But what if we only allow subtractions "at the very beginning," that is, if besides nonnegative real constants and variables $x_1, \ldots, x_n$, the circuits can also use the tropical reciprocals $-x_1, \ldots, -x_n$ of variables as inputs? Let us call such extended: circuits (min, +, $-x_i$) *circuits*. Can reciprocal inputs $-x_i$ substantially reduce the size of (min, +) circuits, that is, can the gap (min, +)/(min, +, $-x_i$) be large?

Let $\mathbb{T}_{\min}[x_1, \ldots, x_n]$ denote the set of all tropical (min, +) polynomials $f(x) = \min_{a \in A}\langle a, x \rangle + c_a$ with $A \subseteq \mathbb{N}^n$ and $c_a \in \mathbb{R}_+$.

**Theorem 6.11 (Jukna, Seiwert and Sergeev [6])** *If a polynomial* $f \in \mathbb{T}_{\min}[x_1, \ldots, x_n]$ *can be computed by a* (min, +, $-x_i$) *circuit of size* $s$, *then* $f$ *can also be computed by a* (min, +) *circuit of size* $\mathcal{O}(ns^2)$.

Recall that the *minimum weight spanning tree problem* (MST problem) on $K_n$ is, given an assignment $x$ of nonnegative weights to the edges of $K_n$, to compute the minimum weight of a spanning tree of $K_n$.

**Corollary 6.12** *The MST problem on* $K_n$ *can be solved by a* (min, max, +) *circuit of size* $\mathcal{O}(n^3)$ *but requires* (min, +, $-x_i$) *circuits of size* $2^{\Omega(\sqrt{n})}$.

**Proof** That this problem can be solved by a (min, max, +) circuit of size $\mathcal{O}(n^3)$ is shown in Theorem 6.5. On the other hand, Theorem 3.16 shows that (min, +) circuits solving this problem require size $2^{\Omega(\sqrt{n})}$, and Theorem 6.11 implies that (min, +, $-x_i$) circuits also require size $2^{\Omega(\sqrt{n})}$ to solve this problem.                    □

**Remark 6.13 (Domain Matters)** As shown by Razborov [13] and Tardos [15], the Boolean gap $(\vee, \wedge)/(\vee, \wedge, \neg)$ can be even exponential. Note that, when restricted to the Boolean domain $\{0, 1\}$, tropical (min, $+$, $-x_i$) circuits are extremely powerful and have the full power of Boolean $(\vee, \wedge, \neg)$ circuits because then $x \wedge y = \min\{x, y\}$ and $x \vee y = \min\{1, x + y\}$ and $\neg x_i = 1 - x_i = 1 + (-x_i)$ for all variables $x_i$. But Theorem 6.11 shows that the gap (min, $+$)/(min, $+$, $-x_i$) is always at most quadratic. The point is that unlike Boolean circuits, tropical circuits are required to correctly compute their values on *all* input weightings $x \in \mathbb{R}^n_+$, not only on 0-1 weightings $x \in \{0, 1\}^n$. Also, when restricted to the Boolean domain $\{0, 1\}$, the MST problem on $K_n$ turns into the graph connectivity problem $\text{CONN}_n$: every assignment $x$ of 0/1 values to the edges of $K_n$ specifies a subgraph $G_x$ of $K_n$, and $\text{CONN}_n(x) = 1$ iff the graph $G_x$ is connected. The Boolean function $\text{CONN}_n$ can be easily computed by a monotone Boolean circuit of size $\mathcal{O}(n^3)$ using, say, the Boolean $(\vee, \wedge)$ version of the Floyd–Warshall–Roy all-pairs shortest path DP algorithm (see Example 1.8). But Corollary 6.12 shows that the MST problem itself requires (min, $+$, $-x_i$) circuits of size $2^{\Omega(\sqrt{n})}$. $\qquad\square$

Theorem 6.11 also yields the following fact about the tropical circuit complexity of dual tropical polynomials. The *dual* of a (max, $+$) polynomial $f(x) = \max_{a \in A} \langle a, x \rangle$ with $A \subseteq \{0, 1\}^n$ is the (min, $+$) polynomial $f^*(x) := \min_{a \in \text{co-}A} \langle a, x \rangle$, where $\text{co-}A := \{\vec{1} - a : a \in A\}$ is the set of complementary vectors.

**Corollary 6.14** *Let $A \subseteq \{0, 1\}^n$ be an antichain. If a (max, $+$) polynomial $f(x) = \max_{a \in A} \langle a, x \rangle$ can be computed by a (max, $+$) circuit of size $s$, then its dual $f^*(x) = \min_{a \in A} \langle \vec{1} - a, x \rangle$ can be computed by a (min, $+$) circuit of size $\mathcal{O}(ns^2 + n^3)$.*

**Proof** Let $\Phi$ be a (max, $+$) circuit of size $s$ computing $f$, and let $g(x) = \max_{b \in B} \langle b, x \rangle + c_b$ be the (max, $+$) polynomial produced by $\Phi$. Since $g(\vec{0}) = f(\vec{0})$ must hold, we have $c_b = 0$ for all $b \in B$. By Lemma 1.24, we know that the inclusions $A \subseteq B \subseteq A^\downarrow$ hold. In particular, the set $B$ also consists of only 0-1 vectors.

Turn the circuit $\Phi$ into a (min, $+$, $-x_i$) circuit: replace every max gate by a min gate and every input variable $x_i$ by its reciprocal $-x_i$. The resulting (min, $+$, $-x_i$) circuit $\Phi'$ produces the Laurent[2] (min, $+$) polynomial $g'(x) = \min_{b \in B} \langle b, -x \rangle = \min_{b \in B} \langle -b, x \rangle$. Then the (min, $+$, $-x_i$) circuit $\Phi''(x) = x_1 + \cdots + x_n + \Phi'(x)$ of size $n + s$ produces and, hence, also computes the (non-Laurent) polynomial $\langle \vec{1}, x \rangle + \min_{b \in B} \langle -b, x \rangle = \min_{b \in B} \langle \vec{1} - b, x \rangle = \min_{b \in \text{co-}B} \langle b, x \rangle = g^*(x)$ (the dual of the polynomial $g$). Then, by Theorem 6.11, the polynomial $g^*(x)$ can also be computed by a (min, $+$) circuit of size $\mathcal{O}(ns^2 + n^3)$, and it remains to show that the polynomial $g^*(x)$ is equivalent to the dual $f^*$ of our polynomial $f$, i.e., that $g^*(x) = f^*(x)$ for holds for all $x \in \mathbb{R}^n_+$. To show this, note that the inclusions $A \subseteq B \subseteq A^\downarrow$ imply the inclusions $\text{co-}A \subseteq \text{co-}B \subseteq (\text{co-}A)^\uparrow$ (because $b \leqslant a$ iff $\vec{1} - b \geqslant \vec{1} - a$). Moreover, since the set $A$ is an antichain, the set $\text{co-}A$ is

---

[2] An arithmetic polynomial $P$ is a *Laurent* polynomial if it can also contain *negative* (integer) exponents. By analogy, tropical *Laurent* (min, $+$) polynomials have the form $f(x) = \min_{a \in A} \langle a, x \rangle + c_a$, for some set $A \subseteq \mathbb{Z}^n$ of "exponent" vectors (and some "coefficients" $c_a \in \mathbb{R}_+$), that is, now the inclusion $A \subseteq \mathbb{N}^n$ does not need to hold.

also an antichain. Thus, by Lemma 1.22, the polynomial $g^*$ is equivalent to $f^*$, as desired.                                                                                                   □

**Proof of Theorem 6.11** We will obtain Theorem 6.11 as an easy consequence of the following two fairly simple lemmas (Lemmas 6.15 and 6.16). Since in $(\min, +, -x_i)$ circuits, we have reciprocal inputs $-x_1, \ldots, -x_n$, polynomials produced by such circuits are, in general, Laurent polynomials.

Our starting point is the following lemma showing that tropical Laurent polynomials produced by $(\min, +, -x_i)$ circuits are of a very special form "a tropical polynomial minus a tropical monomial." Recall that $\mathbb{T}_{\min}[x_1, \ldots, x_n]$ denotes the set of all (non-Laurent) tropical $n$-variate $(\min, +)$ polynomials $f(x) = \min_{a \in A} \langle a, x \rangle + c_a$ with $A \subseteq \mathbb{N}^n$ and $c_a \in \mathbb{R}_+$.                                                                                         □

**Lemma 6.15** *Let $\Phi$ be a $(\min, +, -x_i)$ circuit of size $s$, and let $h$ be a tropical Laurent polynomial produced by $\Phi$. Then there exist a polynomial $g \in \mathbb{T}_{\min}[x_1, \ldots, x_n]$ and a vector $v \in \mathbb{N}^n$ such that $h = g - \langle v, x \rangle$, and both $g$ and $\langle v, x \rangle$ can be simultaneously produced by a $(\min, +)$ circuit $\Phi'$ of size at most $4s$.*

**Proof** We are going to build the $(\min, +)$ circuit $\Phi'$ by traversing the circuit $\Phi$ from inputs toward outputs. At the inputs $x_i$, $-x_i$, and $c \in \mathbb{R}_+$, the corresponding pairs $(g, v)$ with $h = g - \langle v, x \rangle$ are $(x_i, \vec{0})$, $(0, \vec{e}_i)$, and $(c, \vec{0})$.

Now assume that the pairs $(g_1, v_1)$ and $(g_2, v_2)$ at the predecessors of some gate with $h_1 = g_1 - \langle v_1, x \rangle$ and $h_2 = g_2 - \langle v_2, x \rangle$ are already produced. We have to show that the tropical Laurent polynomial produced at the gate itself is also of the form $h = g - \langle v, x \rangle$, where $g$ is a polynomial and $v \in \mathbb{N}^n$. If this is an addition gate, then $h = h_1 + h_2 = (g_1 - \langle v_1, x \rangle) + (g_2 - \langle v_2, x \rangle) = (g_1 + g_2) - \langle v_1 + v_2, x \rangle$, and we can take $g = g_1 + g_2$ and $\langle v, x \rangle = \langle v_1, x \rangle + \langle v_2, x \rangle$; we replaced one addition gate by two addition gates in this case. If this is a min gate, then $h = \min\{h_1, h_2\} = \min\{g_1 - \langle v_1, x \rangle, g_2 - \langle v_2, x \rangle\} = \min\{g_1 + \langle v_2, x \rangle, g_2 + \langle v_1, x \rangle\} - (\langle v_1, x \rangle + \langle v_2, x \rangle)$, and we can take $g = \min\{g_1 + \langle v_2, x \rangle, g_2 + \langle v_1, x \rangle\}$ and $\langle v, x \rangle = \langle v_1, x \rangle + \langle v_2, x \rangle$. In this case, we replaced one min gate by one min gate and three addition gates. The resulting $(\min, +)$ circuit has at most four times more gates than the original $(\min, +, -x_i)$ circuit $\Phi$, as claimed.                                                              □

Lemma 6.15 "replaces" reciprocal inputs $-x_1, \ldots, -x_n$ by one subtraction gate at the very end. To remove this subtraction, we will use the concepts of "greatest common divisors" and "contractions" of tropical polynomials.

A *common divisor* of a polynomial $g(x) = \min_{a \in A} \langle a, x \rangle + c_a$ with $A \subseteq \mathbb{N}^n$ is any vector $v \in \mathbb{N}^n$ such that $a - v \geqslant \vec{0}$ holds for all $a \in A$, and the *greatest common divisor*[3] of $g$, denoted by $\gcd(g)$, is the vector $w = (w_1, \ldots, w_n)$ with $w_i = \min\{a_i : a \in A\}$. That is, $\gcd(g)$ is the largest vector $w \in \mathbb{N}^n$ for which $A - w \subseteq \mathbb{N}^n$ holds (no negative positions). For example, if for every position $i \in [n]$

---

[3] The terminology is by analogy with arithmetic polynomials. A monomial $X^v = \prod_{i=1}^n x_i^{v_i}$ divides a monomial $X^a$ iff $a - v \geqslant \vec{0}$, and the greatest common divisor of the monomials of a given polynomial $P$ is a monomial $X^w$ of largest degree $w_1 + w_2 + \cdots + w_n$ dividing all monomial of $P$.

there is a vector $a \in A$ with $a_i = 0$, then $\gcd(f) = \vec{0}$. The *contraction* of the polynomial $g$ is the tropical polynomial

$$[g](x) := g(x) - \langle w, x \rangle = \min_{a \in A} \langle a - w, x \rangle + c_a,$$

where $w = \gcd(g)$. Note that $g(x) = [g](x) + \langle w, x \rangle$ (addition instead of subtraction).

**Lemma 6.16** *If a polynomial $g \in \mathbb{T}_{min}[x_1, \ldots, x_n]$ can be produced by a* (min, +) *circuit of size $s$, then its contraction $[g]$ can be produced by a* (min, +) *circuit of size $\mathcal{O}(ns^2)$.*

**Proof** Let $\Phi$ be a (min, +) circuit of size $s$ producing the (min, +) polynomial $g$. Call a vector $v \in \mathbb{N}^n$ *$d$-bounded* if $v_i \leqslant d$ for all $i = 1, \ldots, n$. A tropical monomial $\langle v, x \rangle$ is *$d$-bounded* if the vector $v$ is $d$-bounded. In a *$d$-extended* (min, +) circuit, any $d$-bounded tropical monomials $\langle v, x \rangle$ can be used as inputs (for free). □

**Claim 6.17** *The contraction $[g]$ of $g$ can be produced by a $2^s$-extended* (min, +) *circuit of size at most $3s$.*

**Proof** Again, we are going to build the $2^s$-extended (min, +) circuit $\Phi'$ producing $[g]$ by traversing the circuit $\Phi$ (producing $g$) from inputs toward outputs. If $g = x_i$ is an input variable, then $\gcd(g) = \vec{e}_i$ because then $g = \langle \vec{e}_i, x \rangle$; hence, $[g] = \langle \vec{e}_i, x \rangle - \langle \gcd(g), x \rangle = 0$. If $g = c \in \mathbb{R}_+$ is an input constant, then $\gcd(g) = \vec{0}$ because $g = \langle \vec{0}, x \rangle + c$; hence, $[g] = \langle \vec{0}, x \rangle + c - \langle \gcd(g), x \rangle = c$.

Now assume that we are dealing with some gate in $\Phi$ producing a polynomial $g$ from the polynomials $g_1$ and $g_2$ produced at its predecessors. By construction, we have already built a part of $\Phi'$ producing the contractions $[g_1] = g_1 - \langle w_1, x \rangle$ and $[g_2] = g_2 - \langle w_2, x \rangle$, where $w_1 = \gcd(g_1)$ and $w_2 = \gcd(g_2)$ are the greatest common divisors of the polynomials $g_1$ and $g_2$. Let $w = \gcd(g)$ be the greatest common divisor of the polynomial $g$. If $g = g_1 + g_2$, then $w = w_1 + w_2$. Hence, in this case, we have $[g] = g - \langle w, x \rangle = (g_1 - \langle w_1, x \rangle) + (g_2 - \langle w_2, x \rangle) = [g_1] + [g_2]$, and we need no new gates in this case. If $g = \min\{g_1, g_2\}$, then the set of "exponent" vectors of $g$ is the union of the sets of "exponent" vectors of $g_1$ and $g_2$. This, in particular, yields $w \leqslant w_1$ and $w \leqslant w_2$ in this case. Hence, both vectors $w_1 - w$ and $w_2 - w$ are nonnegative, and we can produce the contraction $[g]$ of $g$ as $[g] = g - \langle w, x \rangle = \min\{[g_1] + \langle w_1, x \rangle, [g_2] + \langle w_2, x \rangle\} - \langle w, x \rangle = \min\{[g_1] + \langle w_1 - w, x \rangle, [g_2] + \langle w_2 - w, x \rangle\}$. We replaced one min gate by two addition gates and one min gate. The resulting extended (min, +) circuit $\Phi'$ has at most $3s$ gates and produces the contraction $[g]$ of the polynomial $g$. Since the original circuit $\Phi$ producing the polynomial $g$ had only $s$ gates, and since gates have fan-in two, the greatest common divisors $w$ of the polynomials produced at its gates are $2^s$-bounded vectors, and we can use the corresponding tropical monomials $\langle w, x \rangle$ as inputs for free. □

**Claim 6.18** *Every $d$-bounded tropical monomial $\langle w, x \rangle = w_1 x_1 + \cdots + w_n x_n$ can be produced using $\mathcal{O}(n \log d)$ addition gates.*

**Proof** Each term $w_i x_i$ is the sum $x_i + x_i + \cdots + x_i$ of the $i$th variable $w_i$ times. Let $k := \lceil \log d \rceil$. By repeated "squaring," for each variable $x_i$, all sums $2x_i = x_i + x_i, 2^2 x_i = 2x_i + 2x_i, \ldots, 2^k x_i = 2^{k-1} x_i + 2^{k-1} x_i$ can be simultaneously produced using only $k$ addition gates. Then each sum $w_i x_i$ with $w_i \leqslant d$ (integer $w_i$ being not necessarily a *power* of 2) can be produced using at most $k$ extra additions (by looking at the binary expansion of $w_i$). Thus, $\langle w, x \rangle = w_1 x_1 + \cdots + w_n x_n$ can be produced using at most $2nk + (n-1) = \mathcal{O}(n \log d)$ addition gates.                        □

The $2^s$-extended $(\min, +)$ circuit guaranteed by Claim 6.17 has at most $3s$ gates and, hence, has at most $6s$ $2^s$-bounded tropical monomials $\langle w, x \rangle$ as inputs. By Claim 6.18, each of these monomials $\langle w, x \rangle$ can be produced using $\mathcal{O}(ns)$ addition gates. Thus, we obtain a desired $(\min, +)$ circuit (conventional, with "nothing for free") that produces the contraction $[g]$ by using at most $\mathcal{O}(ns^2)$ gates.          □

**Actual Proof of Theorem 6.11** Suppose that a polynomial $f \in \mathbb{T}_{\min}[x_1, \ldots, x_n]$ can be computed by a $(\min, +, -x_i)$ circuit $\Phi$ of size $s$. Our goal is to show that then $f$ can also be computed by a $(\min, +)$ circuit of size $\mathcal{O}(ns^2)$. Let $h$ be the tropical Laurent polynomial produced by the circuit $\Phi$. Since this circuit computes our polynomial $f$, we know that the produced polynomial $h$ is equivalent to $f$, that is, $f(x) = h(x)$ holds for all input weightings $x \in \mathbb{R}^n_+$.

By Lemma 6.15, there exist a polynomial $g \in \mathbb{T}_{\min}[x_1, \ldots, x_n]$ and a vector $v \in \mathbb{N}^n$ such that $h = g - \langle v, x \rangle$, and $g$ can be produced by a $(\min, +)$ circuit of size at most $4s$. The polynomial $g$ is of the form $g(x) = \min_{b \in B} \langle b, x \rangle + c_b$ for some set $B \subseteq \mathbb{N}^n$ of "exponent" vectors and "coefficients $c_b \in \mathbb{R}_+$. Hence, $h(x) = \min_{b \in B} \langle b - v, x \rangle + c_b$.

Let us first show that the produced polynomial $h$ is a non-Laurent polynomial, that is, has no negative "exponents." Assume for the sake of contradiction that $b_i - v_i < 0$, that is, $b_i - v_i \leqslant -1$ holds for some vector $b \in B$ and some position $i$. Let $K = 1 + c_b$, where $c_b$ is the "coefficient" of the corresponding to vector $b$ tropical term $\langle b - v, x \rangle + c_b$ of the polynomial $h$, and consider the weighting $x \in \{0, K\}^n$ which sets $x_i := K$ and $x_j := 0$ for all $j \neq i$. On this weighting, we have $f(x) \geqslant 0$ but $h(x) \leqslant \langle b - v, x \rangle + c_b = (b_i - v_i) \cdot K + c_b \leqslant -K + c_b \leqslant -1$, a contradiction with $f(x) = h(x)$.

Thus, the polynomial $h(x) = g(x) - \langle v, x \rangle$ produced by the $(\min, +, -x_i)$ circuit $\Phi$ belongs to $\mathbb{T}_{\min}[x_1, \ldots, x_n]$. Note that the greatest common divisor of the (non-Laurent) polynomial $h$ is of the form $\gcd(h) = \gcd(g) - v$. Since the polynomial $g$ can be produced by a $(\min, +)$ circuit of size at most $4s$, and since gates have fan-in-2, every divisor of the polynomial $g$, including $\gcd(g)$, is $d$-bounded for $d := 2^{4s}$ and, hence, the vector $\gcd(h) = \gcd(g) - v$ is also $d$-bounded. Thus, by Claim 6.18, the tropical monomial $\langle \gcd(h), x \rangle$ can be computed by a $(\min, +)$ circuit $\Phi_1$ of size $\mathcal{O}(n \log d) = \mathcal{O}(ns)$.

By Lemma 6.16, the polynomial $[g]$ can be produced by a $(\min, +)$ circuit $\Phi_2$ of size $\mathcal{O}(ns^2)$. On the other hand, since $g(x) = h(x) + \langle v, x \rangle$, we also have $[g] = [h + \langle v, x \rangle] = [h]$. Thus, since $h = \langle \gcd(h), x \rangle + [h]$, the $(\min, +)$ circuit $\Phi_1 + \Phi_2$ of size $\mathcal{O}(ns) + \mathcal{O}(ns^2) = \mathcal{O}(ns^2)$ produces the polynomial $h$. Since $h$ is equivalent to $f$, the circuit $\Phi$ computes our polynomial $f$, as desired.          □

**Remark 6.19 (Arithmetic Circuits)** When translated from the tropical $(\min, +)$ semiring into the arithmetic $(+, \times)$ semiring, the same argument as used in the proof of Theorem 6.11 yields that every monotone arithmetic $(+, \times, 1/x_i)$ circuit of size $s$ (a $(+, \times)$ circuit with additional arithmetical reciprocals $1/x_1, \ldots, 1/x_n$ of variables) can be simulated by a monotone arithmetic $(+, \times)$ circuit of size $\mathcal{O}(ns^2)$.

**Remark 6.20 (Reciprocals in Maximization)** The situation with $(\max, +, -x_i)$ circuits computing (non-Laurent) $(\max, +)$ polynomials $f(x) = \max_{a \in A} \langle a, x \rangle + c_a$ with $A \subseteq \mathbb{N}^n$ and $c_a \in \mathbb{R}_+$ turned out to be more subtle: in this case, the Laurent polynomials $h$ produced by $(\max, +, -x_i)$ circuits computing the non-Laurent polynomial $f$ may contain negative "exponents" and, hence, cannot be produced by $(\max, +)$ circuits at all. Still, it is shown in [6] that the same upper bound $\mathcal{O}(ns^2)$ as in Theorem 6.11 holds also for $(\max, +)$ polynomials $f$ as long as they are *homogeneous*, that is, if the degrees $a_1 + \cdots + a_n$ of all vectors $a \in A$ are the same. Whether this also holds for non-homogeneous $(\max, +)$ polynomials $f$ remains unclear.                                                                           □

## 6.4   Notes on Tropical (min, +, −) Circuits

As before, tropical $(\min, +, -)$ circuits are $(\min, +)$ circuits that can use subtraction gates $(-)$. Let us first show that $(\min, +, -)$ circuits are essentially $(\min, \max, +, -x_i)$ circuits, that is, $(\min, \max, +)$ circuits with reciprocal inputs $-x_1, \ldots, -x_n$. That is, allowing subtraction gates in $(\min, +)$ circuits is essentially the same as allowing reciprocal inputs and max gates.

**Proposition 6.21** *Every $(\min, \max, +, -x_i)$ circuit can be simulated by at most four times larger $(\min, +, -)$ circuit, and every $(\min, +, -)$ circuit can be simulated by at most two times larger $(\min, \max, +, -x_i)$ circuit.*

**Proof** Due to the equality $\max\{a, b\} = -\min\{-a, -b\}$, $(\min, \max, +, -x_i)$ circuits can be simulated by at most four times larger $(\min, +, -)$ circuits.

Now take an arbitrary $(\min, +, -)$ circuit $\Phi$ of size $s$. Our goal is to move all subtractions toward the input variables. To do this, we replace each gate $g$ by two gates $g$ and $-g$. In particular, if $g = x_i$ is an input variable, then we have two input gates $g = x_i$ and $-g = -x_i$ in the new circuit $\Phi'$. If $a$ and $b$ are the gates entering a non-input gate $g$ in $\Phi$, then wire these gates in $\Phi'$ according to the following obvious rules. If $g = a + b$ in $\Phi$, then $g = a + b$ and $-g = (-a) + (-b)$ in $\Phi'$. If $g = \min\{a, b\}$ in $\Phi$, then $g = \min\{a, b\}$ and $-g = \max\{-a, -b\}$ in $\Phi'$. If $g = a - b$ in $\Phi$, then $g = a + (-b)$ and $-g = (-a) + b$ in $\Phi'$. Thus, the new $(\min, \max, +, -x_i)$ circuit $\Phi'$ uses only gates from $\{\min, \max, +\}$, has size at most $2s$, and computes the same function.                                           □

Let us now show that every $(\min, +, -)$ circuit is essentially a difference of two $(\min, +)$ circuits. That is, all subtraction $(-)$ gates in a $(\min, +, -)$ circuit can be moved to just one subtraction gate at the "very end," at the output gate.

**Proposition 6.22** *Every* (min, +, −) *circuit of size s can be transformed into an equivalent* (min, +, −) *circuit of size at most 2s and of the form* $\Phi = \Phi_1 - \Phi_2$, *where both* $\Phi_1$ *and* $\Phi_2$ *are* (min, +) *circuits.*

**Proof** We can move all subtractions (−) to the output gate using the equations: $(a-b)+c = (a+c)-b, (a-b)-c = a-(b+c), \min\{a-b, c\} = \min\{a, b+c\}-b$.          □

Recall that a tropical (min, +) polynomial $f(x) = \min_{a \in A} \langle a, x \rangle + c_a$ is *constant-free* if $c_a = 0$ holds for all $a \in A$. A tropical (min, +, −) circuit is *constant-free* if it uses no nonzero constants as inputs. The following lemma extends Lemma 1.29 to (min, +, −) circuits.

**Lemma 6.23 (Eliminating Constant Inputs)** *If a constant-free* (min, +) *polynomial f can be computed by a tropical* (min, +, −) *circuit of size s, then f can also be computed by a constant-free* (min, +, −) *circuit of size 2s.*

**Proof** Let $f(x) = \min_{a \in A} \langle a, x \rangle + \alpha_a$ be a constant-free (min, +) polynomial; hence, $A \subseteq \mathbb{N}^n$ and $\alpha_a = 0$ for all $a \in A$. Note that this polynomial is *linearly homogeneous* in that $f(\lambda \cdot x) = \lambda \cdot f(x)$ holds for all $\lambda \in \mathbb{R}_+$ and $x \in \mathbb{R}_+^n$. Suppose that $f$ can be computed by a tropical (min, +, −) circuit $\Phi$ of size $s$. By Proposition 6.22, the circuit can be transformed into the (min, +, −) circuit $\Phi = \Phi_1 - \Phi_2$ of size $2s$, where $\Phi_1$ and $\Phi_2$ are (min, +) circuits. Thus, if $g_1(x) = \min_{b \in B} \langle b, x \rangle + \alpha_b$ and $g_2(x) = \min_{c \in C} \langle c, x \rangle + \alpha_c$ are the (min, +) polynomials produced by the circuits $\Phi_1$ and $\Phi_2$, where $\alpha_b, \alpha_c \in \mathbb{R}_+$ are the "coefficients" of these polynomials, then the polynomial produced by the circuit $\Phi$ is of the form $g(x) = g_1(x) - g_2(x)$. Since the circuit $\Phi$ computes the polynomial $f$, we have $g(x) = f(x)$ for all input weightings $x \in \mathbb{R}_+^n$.

Replace all constant inputs of the circuit $\Phi$ by constant 0, and let $\Phi^o$ be the resulting constant-free (min, +, −) circuit (recall that we allow tropical constant-free circuits to use constant 0 as an input). Since constant inputs can only affect the "coefficients" $\alpha_b$ and $\alpha_c$ of the polynomials $g_1$ and $g_2$, the circuit $\Phi^o$ produces the constant-free polynomial $g^o = g_1^o - g_2^o$, where $g_1^o(x) = \min_{b \in B} \langle b, x \rangle$ and $g_2^o(x) = \min_{c \in C} \langle c, x \rangle$ are constant-free (min, +) polynomials produced by the constant-free versions $\Phi_1^o$ and $\Phi_2^o$ of the (min, +) circuits $\Phi_1$ and $\Phi_2$. The polynomial $g^o$ is linearly homogeneous because it has no nonzero "coefficients, while the polynomial $g$ is linearly homogeneous because it is equivalent to our linearly homogeneous polynomial $f$. So, the function $= g(x) - g^o(x)$ is also linearly homogeneous.

**Claim 6.24** *Let* $h : \mathbb{R}_+^n \to \mathbb{R}$ *be a linearly homogeneous function. If there is a constant c such that* $|h(x)| \leqslant c$ *holds for all* $x \in \mathbb{R}_+^n$, *then* $h(x) = 0$ *for all* $x \in \mathbb{R}_+^n$.

Indeed, if $h(x) \neq 0$ for some $x \in \mathbb{R}^n$, then $d = |h(x)|$ is positive. But then for every constant $c > 0$, taking $\lambda := 2c/d$, we obtain $|h(\lambda \cdot x)| = \lambda \cdot |h(x)| = \lambda \cdot d = 2c > c$.

To apply Claim 6.24 to the linearly homogeneous function $h(x) := g(x) - g^o(x)$, let $\alpha_B = \max\{\alpha_b : b \in B\}$ and $\alpha_C = \max\{\alpha_c : c \in C\}$ be the largest "coefficients" of the polynomials $g_1$ and $g_2$. Then $g_1^o(x) \leqslant g_1(x) \leqslant g_1^o(x) + \alpha_B$ and $g_2^o(x) \leqslant g_2(x) \leqslant g_2^o(x) + \alpha_C$ holds for all $x \in \mathbb{R}_+^n$. Thus, $|h(x)| \leqslant c$ holds for the constant $c := \alpha_B + \alpha_C$ and all $x \in \mathbb{R}_+^n$. Claim 6.24 implies that $h(x) = 0$ and, hence, also $g^o(x) = g(x)$ holds for all input weightings $x \in \mathbb{R}_+^n$, that is, the constant-free version $\Phi^o$ of the circuit $\Phi$ also computes our polynomial $f$, as desired.          □

The following lemma shows that there *exist* a huge number of homogeneous $0/1$ minimization problems requiring $(\min, +, -)$ circuits of exponential size.

**Lemma 6.25** *Minimization problems on at least* $2^{2^{\Omega(n)}}$ *homogeneous sets* $A \subseteq \{0, 1\}^n$ *require* $(\min, +, -)$ *circuits of size* $2^{\Omega(n)}$.

**Proof** Let $m = n/2$. We use counting argument, similar to that used in the proof of Corollary 4.13. Let $A_1, \ldots, A_N \subseteq \{0, 1\}^n$ be all $N = 2^{\binom{n}{m}} \geqslant 2^{2^{n/2}}$ subsets of the $m$th slice of the cube $\{0, 1\}^n$ (consisting of $\binom{n}{m}$ vectors). Hence, vectors in each set $A_i$ have the same number $m$ of 1s. Let $f_i(x) = \min_{a \in A_i} \langle a, x \rangle$ be the minimization problem on the $i$th set $A_i$. Let us first show that all these minimization problems are *distinct*[4] (as functions $f_i : \mathbb{R}_+^n \to \mathbb{R}_+$). To see this, suppose for a contradiction there are $i \neq j$ such that $f_i(x) = f_j(x)$ holds for all input weightings $x \in \mathbb{R}_+^n$. Then Lemma 1.22 gives the inclusions $A_i \subseteq A_j \subseteq A_i^\uparrow$. But since vectors in $A_i$ and $A_j$ have the *same* number of 1's, $A_j \subseteq A_i^\uparrow$ yields $A_j \subseteq A_i$. Hence, $A_i = A_j$, a contradiction with $i \neq j$.

Now consider minimal $(\min, +, -)$ circuits solving the problems $f_1, \ldots, f_N$. By Lemma 6.23, we can assume that all these circuits are constant-free. The same simple argument as in the proof of Proposition 1.1 shows that there are at most $L(n, t) = (n + t)^{ct}$ constant-free $(\min, +, -)$ circuits of size $\leqslant t$, where $c$ is an absolute constant. For $t := 2^{n/3}$, we have $\log L(n, t) \leqslant c 2^{n/3} \log(n + 2^{n/3}) \ll 2^{n/2} \leqslant \log N$. Thus, since one single $(\min, +, -)$ circuit can solve only one minimization problem, all but a neglected portion of our minimization problems $f_1, \ldots, f_N$ require $(\min, +, -)$ circuits of size at least $t = 2^{n/3}$.          □

## 6.5 Open Problems

We know that if the set $A$ of feasible solutions consists of only 0-1 vectors and is homogeneous, then $\mathrm{Min}(A) = \mathrm{Max}(A)$ holds (Lemma 1.34). That is, then tropical complexity of maximization and minimization problems is the *same*. For non-homogeneous sets $A$, the situation is different. That then gap $\mathrm{Max}(A)/\mathrm{Min}(A)$ can be exponential is demonstrated by the (non-homogeneous) set $A \subseteq \{0, 1\}^{\binom{n}{2}}$ of characteristic 0-1 vectors of all $s$-$t$ paths in $K_n$ (Theorem 3.6). What about the gap $\mathrm{Min}(A)/\mathrm{Max}(A)$? We already know (Corollary 6.14 in Sect. 6.3) that the gaps between the tropical circuit complexities of optimization problems and their

---

[4] The fact the sets $A_i$ consist of only 0-1 vectors is here important, see Remark 1.35.

duals are at most about quadratic: $\mathtt{Min}(A)/\mathtt{Max}(\text{co-}A)^2 = \mathcal{O}(n)$ holds for any set $A \subseteq \{0,1\}^n$, where co-$A := \{\vec{1} - a \colon a \in A\}$. But what about the optimization problems on the *same* set $A$ of feasible solutions?

**Problem 1** *Are there antichains $A \subseteq \mathbb{N}^n$ for which the gap $\mathtt{Min}(A)/\mathtt{Max}(A)$ is super-polynomial?*

Requirement that $A$ must be an antichain is to avoid trivial, artificial gaps. Namely, one could take *any* set $B \subseteq \{0,1\}^n$ for which $\mathtt{Min}(B)$ is large and consider the set $A = B \cup \{\vec{1}\}$. By the antichain lemma (Lemma 1.33), $\mathtt{Min}(A)$ remains large because vector $\vec{1}$ does not belong to the lower antichain of $A$, but $\mathtt{Max}(A) \leqslant n$. Also, it is important in Problem 1 that a "small" (max, +) circuit (ensuring that $\mathtt{Max}(A)$ is "small") must solve the maximization problem on $A$ *exactly* (within factor $r = 1$). Namely, by Corollary 4.13, we already know that there exist many (even homogeneous) sets $A \subseteq \{0,1\}^n$ such that $\mathtt{Min}(A)/\mathtt{Max}_r(A) = 2^{\Omega(n)}$ holds already for a factor $r = 1 + o(1)$.

We know that (tropical) reciprocal inputs $-x_1, \ldots, -x_n$ *cannot* substantially speed up (min, +) circuits (Theorem 6.11). On the other hand, we also know that max gates already *can* even exponentially speed up (min, +) circuits (Theorems 3.16 and 6.5). So, a natural question is (see Fig. 6.2): can reciprocal inputs substantially decrease the size of (min, max, +) circuits? Since by Proposition 6.21, the (min, +, −) and (min, max, +, −$x_i$) circuit complexities differ by at most constant factors, this question is equivalent to the following.

**Problem 2** *Can the (min, max, +)/(min, +, −) gap be super-polynomial?*

To resolve Problem 2 in the affirmative, it is enough to find a minimization problem $f$ that can be solved by a "small" (min, +, −) circuit but requires "large" (min, max, +) circuits. By Lemma 4.1, the latter task (lower bound) can be settled by proving a large lower bound on the monotone Boolean circuit complexity of the Boolean version of the minimization problem $f$. So, Problem 2 reduces to showing that some optimization problem, whose Boolean version requires

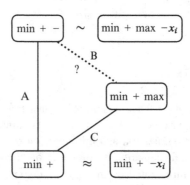

**Fig. 6.2** Basic subclasses of tropical (min, +, −) circuits solving minimization problems. The relation $\sim$ means that the gap can be at most constant (Proposition 6.21), while $\approx$ means that the gap can be at most quadratic (Theorem 6.11). That the gaps (A) and (C) can be exponential is shown by the MST problem (Corollary 6.4 and Theorems 6.5 and 3.16). The status of the gap (B) remains open

"large" monotone Boolean circuits, can be solved by "small" tropical circuits using subtraction gates. A possible candidate could be the *minimum weight perfect matching problem* (known also as the *assignment problem*) $PM_n(x) = \min_M \sum_{e \in M} x_e$, where the minimum is over all perfect matchings $M$ in the complete bipartite $n \times n$ graph. By Corollary 4.3, the problem $PM_n$ requires (min, max, +) circuits of size $n^{\Omega(\log n)}$ even to approximate it within any finite factor. On the other hand, the so-called Hungarian algorithm of Kuhn [8] solves $PM_n$ using a polynomial number of (min, +, −) operations. Unfortunately, this algorithm additionally uses conditional branchings via if-then-else constrains and, hence, cannot be (at least directly) turned into a (min, +, −) circuit.

**Problem 3** *Can $PM_n$ be solved by a* (min, +, −) *circuit of polynomial size?*

A challenging problem is to prove lower bounds on the size of monotone arithmetic (+, ×, /) circuits (monotone arithmetic circuits with division gates) computing an explicit *monotone* polynomial, that is, a polynomial with no negative coefficients.

Even if (+, ×, /) circuits cannot subtract, some non-monotone polynomials (with negative coefficients) are still computable by (+, ×, /) circuits. For example, the polynomial $f(x, y) = x^2 - xy + y^2$ can be computed by a (+, ×, /) circuit $\Phi = (x^3 + y^3)/(x + y)$. A general result of Pólya [11] states that if $f$ is a homogeneous $n$-variate polynomial such that $f(a) > 0$ for all $a \in \mathbb{R}_+^n$ with $a_1 + \cdots + a_n = 1$, then $f(x) = g(x)/(x_1 + \cdots + x_n)^r$ for some $r \geqslant 1$ and some monotone polynomial $g$ (i.e., a polynomial with only positive coefficients). Hence, every polynomial $f$ satisfying Pólya's condition *can* be computed by a (+, ×, /) circuit.

Fomin, Grigoriev, and Koshevoy [3] used this Pólya's result to exhibit a *non-monotone* $n$-variate polynomial $f_n$ (with negative coefficients), which fulfills Pólya's condition and, hence, *can* be computed by a (+, ×, /) circuit as $f_n = g/h$ for monotone polynomials $g$ and $h$ but in any such representation of $f_n$, the polynomial $g$ must have doubly exponential in $n$ degree $d$. Then the desired lower bound on the size $s$ of any (+, ×, /) circuit computing $f_n$ follows from a trivial *degree-bound*: since multiplication (×) gates have fan-in-2, the degree $d$ of the polynomial computed by a (+, ×, /) circuit of size $s$ cannot exceed $2^s$. Thus, any (+, ×, /) circuit computing $f_n$ must have $s \geqslant \log d \geqslant 2^{\Omega(n)}$ gates. Their proof hinges on the fact (proved in [3]) that in any representation of the univariate polynomial $f(x) = (x - 1)^2 + 2^{-2^{n+1}}$ (which trivially fulfills Pólya's condition) as $f = g/h$ for monotone polynomials $g$ and $h$, the polynomial $g$ must have degree at least $2^{2^n}$.

However, no super-polynomial lower bound on the size of (+, ×, /) circuits computing an explicit *monotone* polynomial (without negative coefficients) is known. In the context of tropical (min, +, −) circuits, monotonicity of (arithmetic) polynomials is especially important because only monotone polynomials have their tropical versions: there is no analogue of arithmetic subtraction operation $x - y$ in tropical semirings.

**Problem 4** *Prove a super-polynomial lower bound on the size of monotone arithmetic* (+, ×, /) *circuits computing an explicit monotone polynomial.*

Due to the aforementioned degree-bound, only polynomials $f$ of moderate degree are here interesting. Note that if $f = g/h$ with $h \neq 0$, then $f \cdot h = g$, that is, the polynomial $f$ divides the polynomial $g$. Thus, in view of Remark 6.3, Problem 4 actually asks to exhibit a monotone polynomial $f$ such that monotone arithmetic $(+, \times)$ circuit complexity $\mathrm{Arith}(g)$ is large for *every* monotone polynomial $g \neq 0$ divisible by $f$. That $\mathrm{Arith}(g)$ *can* be much (even exponentially) smaller than $\mathrm{Arith}(f)$ is demonstrated by the (monotone) spanning tree polynomial $f = \mathrm{ST}_n$: then $\mathrm{Arith}(f) = 2^{\Omega(\sqrt{n})}$ (Theorem 3.16), but there is a monotone polynomial $g \neq 0$ such that $f$ divides $g$ and $\mathrm{Arith}(g) = \mathcal{O}(n^3)$ (Theorem 6.1 and Remark 6.3).

A notable progress toward Problem 4 was recently made by Hrubeš and Yehudayoff [4]. Using so-called shadows of Newton polytopes of polynomials, they have already solved Problem 4 for $(+, \times, /)$ *formulas* (circuits whose underlying graphs are trees): every $(+, \times, /)$ formula computing the monotone polynomial

$$\mathrm{SQ}_n(x) = \sum_{\emptyset \neq S \subseteq [n]} \prod_{i, j \in S} x_{i,j}$$

(which we already considered in Corollary 2.5) must have exponential size $2^{\Omega(n)}$. The case of $(+, \times, /)$ *circuits* remains open.

The tropical $(\min, +)$ version $f(x) = \min_{\emptyset \neq S \subseteq [n]} \sum_{i, j \in S} x_{i,j}$ of the polynomial $\mathrm{SQ}_n$ can be computed by a $(\min, +)$ *formula* $\Phi(x) = \min\{x_{i,j} : i, j \in [n]\}$ of size $\mathcal{O}(n^2)$ (even without subtraction gates). Thus, the next challenge—right at the border of our current understanding of the power and weakness of tropical circuits—is to prove non-trivial lower bounds for *tropical* $(\min, +, -)$ circuits or, at least, formulas. Again, due to the trivial "degree bound," only tropical polynomials of moderate degree are here interesting.

**Problem 5** *Prove a super-polynomial lower bound on the size of tropical* $(\min, +, -)$ *circuits or at least for* $(\min, +, -)$ *formulas computing an explicit tropical* $(\min, +)$ *polynomial.*

We already know (Lemma 6.25) that a huge number of $0/1$ minimization problems require $(\min, +, -)$ circuits of exponential size. The problem, however, is to prove such a lower bound for an *explicit* minimization problem. An interesting first progress toward Problem 5 was made by Mahajan, Nimbhorkar, and Tawari [10]. The maximum function $f(x) = \max\{x_1, \ldots, x_n\}$ can be computed as $f(x) = \sum_{i=1}^{n} x_i - \mathrm{MinS}_{n-1}^{n}(x)$, where $\mathrm{MinS}_k^n(x) = \min\{\sum_{i \in S} x_i : S \subseteq [n], |S| = k\}$. It is first mentioned in [10] that $\mathrm{MinS}_{n-1}^n$ can be computed by a $(\min, +)$ formula of size $n\lceil \log n \rceil$ (this is a folklore observation in the context of Boolean threshold functions), and then it is shown that in any representation $f(x) = \Phi_1(x) - \Phi_2(x)$ of $f$ as a difference of two $(\min, +)$ formulas, the formula $\Phi_2$ must have size at least $n \log n$.

As Proposition 6.22 shows, by only doubling the number of gates, every $(\min, +, -)$ *circuit* can be indeed transformed into an equivalent not much larger $(\min, +, -)$ circuit of the form $\Phi(x) = \Phi_1(x) - \Phi_2(x)$, where $\Phi_1$ and $\Phi_2$ are (min, $+$) circuits. Note, however, that when applied to (min, $+$, $-$) *formulas*, such a transformation as in Proposition 6.22 may exponentially explode the formula size: when moving subtractions through min gates via $\min\{a-b, c\} = \min\{a, b+c\} - b$, the entire sub-formula $b$ is duplicated.

# References

1. Cayley, A.: A theorem on trees. Quart. J. Pure Appl. Math. **23**, 376–378 (1889)
2. Chen, W.K.: Graph Theory and Its Engineering Applications. World Scientific, Singapore (1997)
3. Fomin, S., Grigoriev, D., Koshevoy, G.: Subtraction-free complexity, cluster transformations, and spanning trees. Found. Comput. Math. **15**, 1–31 (2016)
4. Hrubeš, P., Yehudayoff, A.: Shadows of Newton polytopes. In: 36th Computational Complexity Conference (CCC 2021), *Leibniz International Proceeding in Informatics*, vol. 200, pp. 9:1–9:23 (2021)
5. Jukna, S., Seiwert, H.: Sorting can exponentially speed up pure dynamic programming. Inf. Processing Letters **159–160**, 105962 (2020)
6. Jukna, S., Seiwert, H., Sergeev, I.: Reciprocal inputs in monotone arithmetic and in tropical circuits are almost useless. Tech. rep., ECCC Report Nr. 170 (2020)
7. Kirchhoff, G.: Über die Auflösung der Gleichungen, auf welche man bei der Untersuchungen der linearen Verteilung galvanischer Ströme geführt wird. Ann. Phys. Chem. **72**, 497–508 (1847)
8. Kuhn, H.W.: The Hungarian method for the assignment problem. Naval Research Logistic **2**, 83–97 (1955)
9. Maggs, B., Plotkin, S.: Minimum-cost spanning tree as a path-finding problem. Inf. Process. Lett. **26**(6), 291–293 (1988)
10. Mahajan, M., Nimbhorkar, P., Tawari, A.: Computing the maximum using *(min,+)* formulas. In: 42nd International Symposium on Mathematical Foundations of Computer Science (MFCS 2017). *Leibniz Internation Proceeding in Informatics*, vol. 83, pp. 74:1–74:11 (2017)
11. Pólya, G.: Über positive Darstellung von Polynomen. In: Vierteljahrsschrif Naturforsch. Ges. Zürich, vol. 73 (1928)
12. Prüfer, H.: Neuer Beweis eines Satzes über Permutationen. Arch. Math. Phys. **27**, 742–744 (1918)
13. Razborov, A.A.: Lower bounds on monotone complexity of the logical permanent. Math. Notes of the Acad. of Sci. of the USSR **37**(6), 485–493 (1985)
14. Strassen, V.: Vermeidung von Divisionen. J. Reine Angew. Math. **264**, 184–202 (1973)
15. Tardos, É.: The gap between monotone and non-monotone circuit complexity is exponential. Combinatorica **8**(1), 141–142 (1988)

Printed in the United States
by Baker & Taylor Publisher Services